served 1919 - 1939

The Turn of the Wheel 1988
has been published
in a Limited Edition
of which this is

Number 11

A list of original subscribers
is printed at
the back of the book.

THE TURN OF THE WHEEL
The History of the RASC 1919-1939

A practice parade for an exhibition held at Olympia in 1933. Behind is a display of modern maintenance work at that time. The front wheel of a Morris six-wheeler being removed and replaced as an ordinary everyday task.

The Author, Major General P. G. Turpin CB OBE MA FCIT.

THE TURN OF THE WHEEL

The History of the RASC 1919-1939

BY

Major General PATRICK G. TURPIN CB OBE MA FCIT

PUBLISHED UNDER THE INSTRUCTIONS OF
THE INSTITUTION OF
THE ROYAL CORPS OF TRANSPORT
BY
BARRACUDA BOOKS LIMITED
BUCKINGHAM, ENGLAND
MCMLXXXVIII

PUBLISHED BY BARRACUDA BOOKS LIMITED
BUCKINGHAM, ENGLAND
AND PRINTED BY
EBENEZER BAYLIS & SON LIMITED
WORCESTER, ENGLAND

BOUND BY
CAMELOT PRESS PLC,
SHIRLEY ROAD, SOUTHAMPTON

JACKET PRINTED BY
CHENEY & SONS LIMITED
BANBURY, OXON

PHOTOLITHOGRAPHY BY
FORE COLOUR GRAPHICS LIMITED
HERTFORD, ENGLAND

DISPLAY SET IN BASKERVILLE
AND TEXT SET IN 10½/12pt BASKERVILLE BY
KEY COMPOSITION
NORTHAMPTON, ENGLAND

ISBN 0 86023 428 2

DEDICATION

Her Royal Highness, Princess Alice,

Duchess of Gloucester GCB CI GCVO GBE

Colonel in Chief

Royal Corps of Transport

FOREWORD

by Field Marshal the Lord Carver GCB CBE DSO MC

Major General Turpin's History of the Royal Army Service Corps between 1919 and 1939 provides a valuable addition, not only to the history of the Corps, but also to that of the British Army as a whole. There are two reasons for this: first that it was an important period in the history of the Army, as it saw its transformation, hesitantly and reluctantly in many quarters, from reliance on the horse as a means of transport on the battlefield, and on the railway behind it, to acceptance of the motor vehicle for both: secondly that, as an army cannot fight, or even exist, without supplies and transport, the history of the activities of the Royal Army Service Corps, which provided both, is a history of the activities of the Army as a whole. The titles of the book's chapters clearly illustrate that.

The most significant feature of the period was the lead taken by the Corps in the introduction of mechanical transport into the Army, notably the development of wheeled vehicles capable of moving off roads, pioneering multi-axle drive and pneumatic tyres. Hand in hand with that went a careful attention to detail in training all concerned, from drivers and mechanics to organizers, methods gradually adopted by the whole Army. The widespread use of motor transport naturally led to the prime importance of the provision and handling of supplies of fuel for them. In all these spheres the Royal Army Service Corps, led by some outstanding officers, showed admirable initiative, pointing the way for the rest of the Army, which owed the Corps a considerable debt of gratitude. Without its lead, the Army would have been even less well prepared to face war in 1939 than it was.

Michael Carver
FM

CONTENTS

PREFACE

In the second volume of *The Royal Army Service Corps, a History of Transport and Supply in The British Army*, Colonel R. H. Beadon carried forward the story of the Army Service Corps from the end of the Boer War to the years immediately following the First World War. *The Story of the RASC 1939-45* took up the story from the outbreak of the Second World War.

This account covers the history of the Corps during the period between the two wars.

It is a story of high hopes for peace, of disarmament and retrenchment in the Armed Forces, of disillusion and world recession in the early 1930s, of re-awakening to the new dangers which threatened us and, finally, of a determination to recover lost ground and face the future with resolution.

When at last hostilities came to an end with the Armistice of November 1918, another era in the life of the Corps had passed. The aftermath of the war to end wars was a period of readjustment to peace-time conditions, with all the problems of demobilisation, re-development of industry, more liberal attitudes and a determination to enjoy the bitter fruits of such a hard-won peace. But disillusion was not long in coming. In spite of the good intentions behind the creation of the League of Nations, it was soon clear that a long period of peace was unlikely. The next two decades became a preparation for a resumption of hostilities.

Although expenditure on the Armed Forces was severely restricted while disarmament was the cry, and in the face of widespread pacifism and a spirit of appeasement, there were some who understood what was happening and, in spite of shortages of equipment and manpower, when the crises of 1938 and 1939 were upon us, the training of the Corps for its role in the next war was both realistic and imaginative. As events were to prove later, it was based on sound principles.

And so this account of the Royal Army Service Corps between 1919 and 1939 will be one of initial contraction, of concentrated training on a cadre basis, culminating in an almost feverish attempt to make up for lost time in the preparation for the war which, after 1935, was clearly inevitable.

These twenty years can be conveniently divided into two periods: the first from 1919 to 1929, when the likelihood of major war occurring was extremely remote and the lessons of the 1914-1918 war were being assessed and absorbed, and the new problems presented by mechanisation were being studied. During the second period, from 1930 to 1939, war was in the air, horse transport was a thing of the past and realists were training for a war in which completely mechanised and highly mobile forces would be involved.

Some aspects of the Corps' activities during these years do not fall readily into such convenient compartments. In such cases it has been thought best to cover the subject in a single chapter spanning the whole of both periods. These subjects are included in Part III.

It will be necessary to pick up some of the threads that are woven into the earlier period before 1919, and to follow some of the pattern into the years following 1939. Without such links the continuity of the story would be lost.

A word of explanation is necessary on the subject of rank. The general principle followed is that the rank which the individual held at the time of the occurrence is used. If a high rank was subsequently reached, this is only shown when the name is first mentioned. The full rank and decorations at the time of retirement are shown in the Index.

A great debt of gratitude is due to all those who have helped in the writing of this history, with their reminiscences, their suggestions and their criticism. It is impossible to mention them all by name, but I should particularly like to thank the following for their kind and generous help:

The late Major General Sir Cecil Smith, whose knowledge of the Corps covered a wider span than anyone else, and who must be regarded as one of the greatest authorities on the Royal Army Service Corps; the late Major General H. C. Goodfellow who, by his enthusiasm and encouragement was largely responsible for my undertaking the task of writing this History; the Commander of the Royal Corps of Transport Training Group and all his staff, who have given me access to all their records and archives, in particular Major C. W. P. Coan, the former Editor of the *Waggoner* and Curator of the Regimental Museum; Brigadier D. W. E. Hancox and Brigadier B. G. E. Courtis who, as Regimental Secretary, have given me all the facilities that I have asked for, and the clerical staff of the Regimental Headquarters who have always been most helpful. I should also like to thank Major Generals D. H. W. Buckle, A. F. J. Elmslie and E. H. G. Lonsdale, Brigadiers E. R. Goode and D. J. Sutton, the late Lieutenant Colonel G. R. C. Barnett and the late Lieutenant Colonel J. G. C. Low, all of whom have lent me material, answered my queries and helped me in a number of ways.

I am particularly grateful to Brigadier B. G. E. Courtis and Lieutenant Colonel T. A. Danton-Rees for their great help in proof reading and preparing the book for publication.

I am grateful to the Imperial War Museum for the opportunity to reproduce the picture on page 17, to Lord Montagu of Beaulieu for the picture of his father, and to the Institution of the Royal Corps of Transport and those others who have generously contributed photographs, as well as to Mr Chadwick for the map.

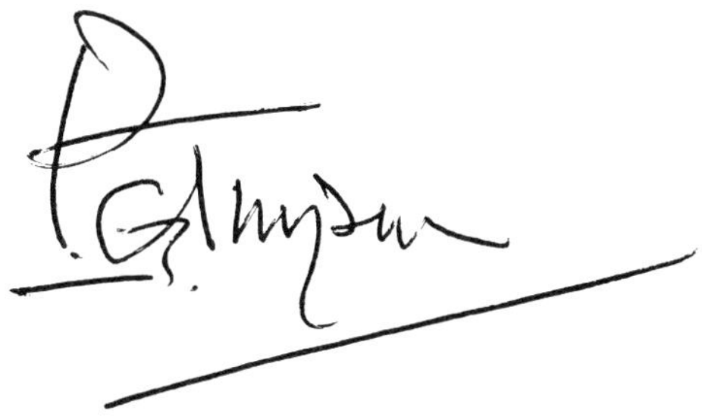

TH/234029 Dr. Oliphant R. M.

ROYAL ARMY SERVICE CORPS.

THE British Armies in France, which once stood in such urgent need of your services, are now able to release you to take your place among your friends at home.

I am very glad that you have come safely through the Campaign. Before you quit the shores of France, I want to take advantage of this opportunity to thank you, both officially as the Director of Transport and personally as one who is proud to have served for 27 years in your Corps, for the really magnificent work which has been done by all ranks of the Transport Branch.

It has not been possible for me to be in such close touch with you as I could have wished—but I know well of the long hours at the reins while well cared for animals plodded through shattered roads, of the wonderful feats of lorry drivers who held the wheel for days and nights without rest, of the artificers on whom all relied—so rarely in vain—and of all the others who played their part in the organization, and I know also that you never failed to "deliver the goods."

I hope that you will meet with the same success at home as has the Royal Army Service Corps in France and Belgium.

Good-bye. A long life and every happiness to you.

B. Boyce

Major-General.
Director of Transport, British Armies in France,

Thank you, driver.

PART I
1919 TO 1929

THE LEAN YEARS — I

'It is the misfortune of our Army that, as soon as it has been brought to high efficiency by a great war, it is instantly pulled to pieces by the nation.' — Sir John Fortescue.

Less than three weeks after the Armistice was signed the Director of Supplies and Transport, Major General Sir Alban Crofton Atkins, issued a Special Corps Memorandum, announcing that 'His Majesty the King has been graciously pleased to confer the title of Royal upon the Army Service Corps in appreciation of the valuable services of this Corps during the war.' In announcing this high honour he conveyed to all ranks of the Corps the congratulations of the Quartermaster General to the Forces, ending with these words: 'In all circumstances the Corps has done its duty, nor can the Quartermaster General recall one single instance where the maintenance of the troops in any part of the world has failed, thus establishing a record of which the whole Army may be justly proud'.

Never before had the Corps been put to such a searching test. The awareness of having been severely tried and not found wanting bestowed a sense of pride and achievement, which left the Corps with a higher state of morale than ever before. This was enhanced by the return of many officers and men, seconded during the latter part of the war to fighting arms — mainly Infantry — who brought back invaluable experience and a wider concept of the Army as a whole. Some 1,200 officers and over 82,000 other ranks were seconded or transferred to the fighting arms.

But this newly won confidence in its ability to carry out its functions under the most trying conditions of war was to be even more sternly tested during the years of peace. The final victory in the 'war to end wars' had left all nations, winners or losers, exhausted. There was a determination that there should be no repetition. The politicians sought, through the terms of the Treaty of Versailles, to prevent any recrudescence of German militarism, and by the establishment of the League of Nations to create an institution which would ensure a lasting peace. The people were sick and tired of war, and only wished to return to civilian occupations, and to enjoy what ephemeral pleasures the twenties could offer in compensation for four and a half years of futile and senseless destruction. Nobody wanted to give much thought to the Armed Services, except those whose duty it was to effect as large economies as possible in public expenditure.

And so the 'Geddes Axe' fell. Ironically enough, the man whose name is associated with the reduction of the Army to little more than cadre strength was Sir Eric Geddes who, as Director General of Movements and Railways in 1916 and 1917, had done so much to expand administrative resources to deal with movement and maintenance during the war.

These drastic measures accorded with the country's mood. The strategic concept, as expressed by the War Cabinet in 1919, was that the 'British Empire will not be engaged in any great war during the next ten years, and that no Expeditionary Force will be required'. This was to be the premise on which the Service departments were to base estimates for the next few years. The assumption was reviewed in 1924 by the Committee of Imperial Defence, and again in 1927 and 1928, when it was re-affirmed that no major war was to be expected for ten years and that this period should be extended continuously, but that the assumption should be reviewed every year by the Committee. Not until 1932, when it was reluctantly accepted that a real threat to peace existed, was that assumption abandoned.

The pillars of our defence strategy were the protection of the sea routes on which our existence as an importing country depended, the maintenance of the free international use of the Suez Canal and the security of Middle East oil. In other words, a policy of Imperial policing, with no plans for military involvement in Europe. A small regular Army and Navy with the help of the newly-formed Royal Air Force would be sufficient, it was assumed, to meet these commitments. In the event of a sudden emergency, it was argued, there were ample reserves of trained officers and men with recent experience of active service conditions, who could be called up at short notice.

Whenever the country embarks on a policy of retrenchment, and heavy reductions are called for in the size of the Army, great weight and influence are applied for the retention of as much as possible of the fighting Arms, at the cost of even greater reductions in the Administrative Services. Although, in peace-time, the former are largely unproductive and can devote most of their time to training, while the Administrative Services are fully occupied in daily maintenance tasks and garrison details, it is argued that many of these duties are of a civilian nature and need not necessarily be performed by military units or uniformed personnel. On the other hand, when the reduced resources of the Royal Army Service Corps prove insufficient and it is necessary to hire civilians, howls of protest are raised by the finance branches.

Between 1914 and 1918 the strength of the Army Service Corps, which in 1914 had around 500 officers and 6,000 other ranks, expanded to nearly 12,000 officers and 325,000 other ranks, a total not even exceeded in the 1939-45 war. During the years immediately following the end of the war the Royal Army Service Corps, in company with the rest of the Army, had to suffer a reduction from this enormous total to one not much larger than its size in 1914 — a process that was no more comfortable than the experience of Alice in Wonderland, when she obeyed the label on the bottle which said 'Drink me'.

But this could not happen all at once. The process of demobilisation, although it was well organised and proceeded smoothly enough, took a long time and had not been completed by the middle of 1920. There were enormous quantities of vehicles and stores to be disposed of. The war had left stricken areas which needed the help of the Allied Armies. There were enemy territories in which occupation forces were required for the maintenance of order until the Peace Treaties were ratified, and in some cases, such as the Mandated Territories, for much longer. And there were still areas where unrest and disturbance persisted and where considerable forces had to be maintained, in particular the Middle East, the North-West Frontier and Ireland.

It is probable that the Corps was more stretched and more widely dispersed throughout the world during the years immediately following the Armistice than at any other time in its existence. Wherever British troops were deployed at home or abroad RASC units or detachments were required to carry out supply, transport and barrack duties. Overseas, peacetime garrisons extended from British Honduras and Jamaica in the west, to Gibraltar, Malta and Cyprus in the Mediterranean, and to Ceylon, Singapore, Hong Kong and North

China in the east. In addition, substantial forces were still maintained in Germany, on the Rhine, in Berlin, in Danzig and in Silesia. Large numbers of troops still occupied Northern Russia and the Black Sea area, and military commitments continued in the Balkans, in Egypt and Palestine, in Mesopotamia and in East and South Africa. In India the Royal Army Service Corps had inherited, since 1916, a heavy commitment, which was to continue until 1928. Finally, industrial unrest, with coal strikes and railway stoppages, created additional burdens.

All these commitments placed an almost intolerable strain on peace-time skeleton forces. As demobilisation continued and the size of the Army was reduced, it became progressively more difficult to find manpower and resources.

One of the immediate effects of retrenchment was to halt promotion prospects. In the Infantry only two majors were allowed on the Peace Establishment of a battalion: the second-in-command and the senior company commander. Even in 1933 there were captains in command of companies, who had been in command in action as majors in the 1914-1918 war, and senior subalterns who had 18 or 19 years' service. The expression 'dead men's shoes' was realistic. Although the situation was a little better in the Cavalry, where a larger number of early retirements could be expected, in the RASC and other Corps, the position was not really improved until reforms were introduced in 1938, when Mr Hore-Belisha (who had served in the Army Service Corps as a major in the 1914-1918 war) became Secretary of State for War. From then on promotion to captain and major was on a time basis, subject, of course, to the appropriate qualifications.

Another blow to the Corps was the withdrawal in July 1919 of 'Corps Pay' from all RASC officers except those with technical qualifications (AMIMechE — Associate Member of the Institute of Mechanical Engineers). Although this was partially restored in April 1925, on a lower scale, for all officers who successfully passed a junior officer's course at the Royal Army Service Corps Training College, that was almost exactly counter-balanced, at least for captains and more junior officers, by reductions in Army pay imposed in October of the same year.

Reduced peace establishments, low rates of pay and pension, poor promotion prospects and lack of new recruits had two serious effects: the difficulty of carrying out adequate and realistic training, and a lowering of the standard of officer and recruit intake. Not until the depression, at the end of the first decade after the war, began to affect employment, did the standard of recruits begin to show a welcome and significant improvement.

This post-war climate was hardly one in which the high morale born during the war could be expected to flourish. Nevertheless, the spirit of the Royal Army Service Corps was not only kept alive but showed a remarkable resilience and inventiveness. When we search for a reason, we find two important factors: first, the high quality of those who were responsible for the higher direction of Corps affairs, and second, the importance that was always placed on the Training Centre, and the care taken in the selection of the officers to command and staff it.

RASC mule teams drawn up at the Base Horse Depôt, North Russia, 1919.

Major General Sir Alban R. Crofton Atkins KCB CMG (Director of Supplies & Transport 1916-1919).

RASC reindeer team, North Russia, 1919. (IWM)

RASC Soldiers in Tientsin, North China, 1920.

HRH the Duke of Connaught and Strathearn, Colonel, later Colonel in Chief, Royal Army Service Corps.

Higher Direction II

HRH the Duke of Connaught and Strathearn had been appointed Colonel of the Army Service Corps in September 1902 and, when the Royal title was bestowed on the Corps in 1919, he became Colonel of the Royal Army Service Corps. In 1932 the title of the appointment was changed to Colonel-in-Chief which he continued to hold until his death on 16 January 1942. Throughout the period with which we are concerned he took a great interest in the affairs of the Corps and gave it his full support.

In the year in which he became Colonel-in-Chief he paid a memorable visit to Buller Barracks at Aldershot and inspected the newly-formed Connaught Squad at drill on 'W' Square.

In 1938, when the Corps was celebrating its 50th Anniversary, he sent the following message which was read out by the Representative Colonel Commandant to the Corps Sunday parade:

> 'I congratulate the Royal Army Service Corps on the celebration of the 50th Anniversary of their formation. The Corps has every reason to feel a sense of pride in its fine record of service during peace and war, and I am proud to be its Colonel-in-Chief.
>
> 'I remember the Corps since its formation and its great-grandfather, the Military Train was commanded by Colonel Clark-Kennedy, who was a personal friend of mine.
>
> 'I hope the Corps may have a very successful celebration, and I only wish I could have taken part in some of the week's events.
>
> *Arthur*, Colonel in Chief.'

He was Patron of the Royal Army Service Corps Officers' Club and the Royal Army Service Corps Regimental Association since their formation. When his many other engagements permitted, he presided at the annual dinner of the Royal Army Service Corps Officers' Club.

There were left, at the end of the war, three 'elder statesmen' who had played leading roles in the formation of the Army Service Corps, and in winning for it recognition as one of the combatant Corps of the Army, with career prospects comparable with those of the fighting arms.

Major General Sir Herbert N. Bunbury KCB was the first officer to join the Army Service Corps from the regimental list on its formation in 1888, and was also the first Army Service Corps officer to be promoted to the rank of major general. In 1894 as a DAQMG (Deputy Assistant Quartermaster General) at the War Office, he signed the order giving approval to the design of the Army Service Corps regimental badge, designed by Captain A. H. Terry ASC. Later he held important staff appointments as Brigadier i/c Administration, Gibraltar, and Major General i/c Administration, Ireland, setting a precedent for future Corps officers. At the end of the war he became President of the Royal Army Service Corps Club, on its revival, and remained so until his death at Mentone in early 1922.

Colonel Sir Edward W. D. Ward Bart GBE KCB KCVO was probably the most distinguished officer to have served in the Corps. He was promoted major in the Army

Service Corps in 1888 on the Supernumerary List (those who had served in the Commissariat and Transport Staff and transferred to the Army Service Corps).

As a brevet-colonel he served in South Africa as AAG (Assistant Adjutant General) at Ladysmith and, later, was specially selected by Lord Roberts to take over all the supply arrangements at Army Headquarters as Director of Supplies. He retired from the Army in 1901 and was appointed Permanent Under-Secretary of State at the War Office; he became the first Secretary of the Army Council in 1904. In this unique position he was able to give great help to the Corps during its development, as he had the ear of both the Military Secretary and the Commander-in-Chief at the War Office. In the same year he was appointed Honorary Colonel of the Canadian Army Service Corps, which had been formed three years earlier in 1901. During the 1914-18 War he was Director-General of Voluntary Organisations and Commander and Chief Staff Officer of the Metropolitan Special Constabulary. After the war he continued to help the Corps with advice and support. He was chairman of the Royal Army Service Corps Club from 1919 until his death in October 1928. When he died, barely four months after presiding at the annual general meeting of the Corps Club, an old brother officer wrote of him:

'No matter how much he was occupied, he always found time to take the deepest interest in everything connected with his old Corps and it was very seldom indeed that he missed attending, and presiding at the many meetings of the Corps Club. No question which affected the welfare of the Corps ever failed to enlist his hearty co-operation, sympathy and wise guidance.'

The third of this trio was Lieutenant General Sir Frederick T. Clayton KCB KCMG who probably did more in the pre-1914 period for the Army Service Corps than anyone else. He transferred to the Permanent List of the Army Service Corps in 1889. After serving in the South African War he was appointed AQMG (Assistant Quartermaster General) for the Army Service Corps at the War Office in 1902, following a period in command of the Corps in Aldershot. Here he was staff adviser to the Quartermaster General for all matters affecting the Army Service Corps. He submitted far-reaching proposals for the improvement of career prospects and sought a fuller recognition of the role of Supplies and Transport in any future war. He was particularly concerned that more vacancies at the Staff College should be available to Army Service Corps officers and that better opportunities should be provided for promotion to the rank of major general. He gave his own personal account of his activities during this period in a lecture to the officers of the Corps on 22 November 1921 (recorded in the *Royal Army Service Corps Quarterly*, April 1922). His next appointment was Director of Supplies at the War Office for three years and he then went as Major General i/c Administration to the South Africa Command. He started the war as Director of Supplies, British Expeditionary Force, and in 1915 was given the important appointment of Inspector General of Communications. When approval was given in August 1921 for the appointment of three Colonels Commandant to the Royal Army Service Corps, he was one of the first three to be selected and held the appointment until the age of 70 in 1925. There was no precise definition of the duties of the Colonels Commandant, but one of their number was to be nominated Representative Colonel Commandant each year, and he would be the equivalent of the Colonel of a Cavalry or Infantry Regiment. Some confusion was caused by the introduction at about the same time of the appointment of Colonel Commandant in place of the rank of Brigadier General, which was abolished in the face of French criticism of the number of generals in the British Army. The appointment of Brigadier was substituted for Colonel Commandant on 1 June 1928, like that of Commodore in the Royal Navy, and eventually became a rank again in 1949.

Sir John Fortescue, in his introduction to Colonel Beadon's Volume II of *The Royal Army Service Corps, History of Supply and Transport in the British Army*, had some hard things to say about the shabby and unworthy treatment of the Corps in the inter-war years. General

Clayton, from his long experience as a champion of the Army Service Corps, had no illusions about the need for the Corps to battle for its rights. The final words of his lecture as the first Representative Colonel Commandant reveal his great fondness for the Corps:

'The Army Service Corps has, during its 32 years of life, had so many disappointments and breaches of faith, that I hope and trust it will not be again necessary for it to have to fight for its rights. It is a Corps, that since its inception, has had nothing but a series of successes, both on service and in peace. It has a record behind it that cannot be beaten by any other Corps in the Army and it deserves a better fate than that of having to fight continuously for its rights. I may be a bit of a pessimist (I hope I am), but I have a great affection for the old Corps. It has done splendidly, and never failed in whatever work it has been called upon to do. I owe everything to it myself; I have never had a more loyal body of officers to work with, and I hope that, throughout the whole of my service in the Corps, I have never made an enemy. I shall always have the interests of the Corps at heart.'

In his retirement he continued to show the liveliest interest in every field of Corps activity and fully lived up to his nickname as 'Father of the Corps'.

The Corps owes a great debt of gratitude to these three officers, who did so much for its good name in the early years.

The RASC was fortunate in that the first three Directors of Supplies and Transport in the post-war period had served in widely different theatres of war: the British Expeditionary Force in France, the East Africa Command and the Egyptian Expeditionary Force in Egypt and Palestine. Their combined experiences covered a great diversity of Royal Army Service Corps activities under varied conditions. There was little chance that Royal Army Service Corps training would be entirely based on the stereotyped maintenance system evolved on the Western Front.

Major General Sir Evan Carter KCMG CB MVO became Director of Supplies and Transport in March 1919 after holding the appointment of Director of Supplies, British Armies in France, from 1915 to the end of the war. Nobody was more knowledgeable than he of the problems of supplying large armies in the static conditions of European trench warfare. But he was the first to realise that the RASC must not be too much influenced by the experiences of his own particular theatre of operations. In his foreword to the first number of the revived *RASC Quarterly* in January 1920 he clearly expressed his views on the course which RASC training should take:

'Now that we *know* our strength, this is the time to consolidate our experiences of the past five years from the military point of view . . . if we have to content ourselves mainly with theory let us make every effort to ensure that our theories of the future are of the best we can produce; in other words, let us take our experiences of the last few years, prune and examine them with open minds, and build our theories for the future on the best foundations we can . . . We must appeal therefore to all officers who served in the RASC during the war to support our efforts in the direction indicated, by giving their comrades the benefit of their experiences and the lessons which they learnt in any and every theatre of war.'

These terms of reference were faithfully followed in subsequent issues of the *Quarterly*, in which accounts by senior officers of their experiences in almost every theatre appeared. It was particularly appropriate that General Carter should have written these words under the heading 'Resurgat', as he had been so closely concerned with the original foundation of the *Quarterly* in 1905, and was its first editor.

During his time as Director of Supplies and Transport, General Carter had to cope with all the problems of demobilisation and readjustment following the end of the war. His chief headache was how to meet all the continuing commitments with rapidly diminishing resources. Even before demobilisation had been completed, it was apparent that the 'axe' had fallen too swiftly and too far. In the Army Estimates for 1920/21 the strength of the Corps had been reduced to 861 officers and 14,231 other ranks: this took little account of the many widespread commitments left after the war, and in the following year's estimates (1921/22) the Corps' strength was increased to 1,221 officers and 16,492 other ranks.

Although so many commitments remained, demobilisation went relentlessly forward and, at times, even with a greatly increased use of civilians, the Corps was hard put to it.

Its future was bound up with the development of mechanical transport, and the replacement of the horse by the motor lorry would be one of the major influences in the RASC during the next few years. In January 1921 a Mechanical Transport Advisory Board was constituted at the War Office under the chairmanship of General Carter. Among its civilian members were Professor W. E. Dalby FRS, Sir Henry Fowler KBE, and Mr H. Niblett CBE. Its terms of reference were not only to act in an advisory capacity on technical problems, but also to secure close liaison with the motor industry, so that the Army would be in touch with the progress of mechanical transport throughout the world. This close liaison continued happily for the next 20 years.

When he retired in August 1921 the pattern of the peace-time RASC was taking shape. Demobilisation was complete; units were reduced to cadre strength (about 25% of war establishment); the replacement of the horse by mechanical transport was well under way; and the Training Centre at Aldershot was preparing for the future.

Just over a month later he was appointed one of the first three Colonels Commandant in company with General Clayton and Major General Sir Frederick Landon KCMG CB. At a farewell dinner given to him on 3 October the menu was decorated with a copy of the well-known cartoon from *Punch* by L. Ravenhill, entitled 'The Sinews of War', in recognition of the part he played as Director of Supplies in France.

Major General P. O. Hazelton CB CMG succeeded General Carter in August 1921 and held the appointment until 1925. Before becoming Director of Supplies and Transport he served for a year in the War Office as AD of Transport, followed by a year as DD of S and T Eastern Command.

When General Hazelton went to East Africa from France in 1916 as Director of Supplies and Transport he found conditions very different from those in Europe. Operations extended over great distances with virtually no developed communications other than the few railway lines. The theatre demanded great powers of improvisation, resourcefulness and the full use of every sort of transport from solid-tyred motor lorries, Ford trucks (with a carrying capacity of 1,000 to 6,000lb), oxen, mules, donkeys, to native carriers. The force was made up of a bewildering number of different races, many of whom had their own particular dietetic requirements. General Smuts never let lack of administrative resources influence him unduly in carrying out his operational plans. He was a hard taskmaster, who took what administrative risks he considered justified in pursuit of his objectives. General Hazelton had plenty of experience of 'making bricks without straw', a useful apprenticeship for the appointment of Director of Supplies and Transport at the War Office in 1921. During his time as Director there was little scope for expansion or development. On the contrary, he had to be content with the limited resources available and to do his best.

Major General G. F. Davies CB CMG CBE spent most of the war in the Middle East, as Director of Supplies and Transport of the Egyptian Expeditionary Force, and was responsible for the maintenance of the large Imperial Forces deployed in that theatre, first, over desert country, in which roads were non-existent, and later, when the breakthrough into Palestine had been made, over hilly and rocky country in one of the swiftest advances in modern warfare. He had the unique experience of having under his control a camel transport corps of unparalleled size. Each camel company consisted of 2,020 camels (each with a load of from 250 to 350 pounds). During the campaign a total of 72,500 camels and 170,000 camel drivers were employed.

In Palestine four donkey companies of 2,000 animals each were used in the Judaean Hills and great use was made of surf boats and sea depôts for coastwise freight maintenance. Like his predecessor, General Davies was no stranger to improvisation.

This tribute was paid by Field-Marshal Lord Allenby to the transport services of the Egyptian Expeditionary Force:

'Seldom, if ever, can the Royal Army Service Corps have been confronted with problems of transport so difficult as those which called for solution during the advance of the Expeditionary Force through Palestine and Syria in 1917-1918.

'Deserts, morasses, hills, valleys, rocks and rugged mountains, all had to be conquered; while climatic conditions included extremes of heat and cold, of drought and torrential rain . . .

'I, who had the honour of commanding the Army, can testify to the colossal nature of the task accomplished; and to the Royal Army Service Corps I owe a deep debt of gratitude.

'No one can fail to be proud of the achievement; a triumph of organisation and perseverance, unsurpassed in the annals of our Empire.'

General Davies succeeded General Hazelton in August 1925, after holding the appointment of Assistant Director and Deputy Director of Transport in the War Office for the first four years after the end of the war, from 1919 to 1923, and then, from April 1923, of Inspector Royal Army Service Corps, with the bizarre rank of colonel-on-the-staff, which had been adopted when the rank of brigadier general was abolished.

The four years of his appointment witnessed the end of the Royal Army Service Corps' involvement in India, the continuation of the eclipse of horse transport and a radical change in the policy for the research and provision of mechanical transport for the Army. At exercises and on collective training, and in studies at the RASC Training College, the organisation of supply and transport services in the field — particularly in the divisional area — was the subject of continuous study and experiment.

However, during this period, theoretical studies at the training college represented the extent of training for war carried out by the Corps. The reduction of units to mere peace-time cadres, the continuing demands for transport to meet unforeseen commitments, disturbances at home and abroad, and the reluctance to resort to the hiring of civilian transport in order to release military units for training, made it almost impossible for RASC units to do more than complete their individual training. This failure to exercise adequately the new intake in its war-time role did not have the serious results which it invited. In the light of the concept 'no war in the next ten years' and the great wealth of surviving experience from the war, the Corps still had the time and means to re-create the same state of efficiency which it reached in the immediate pre-1914 days.

Apart from the three Directors of Supplies and Transport who held office from 1919 to 1929, the Corps had a rich legacy of comparatively young officers who had reached high rank in the war and had held responsible staff, supply and transport appointments in various theatres. During the 1920s, after a temporary reduction in rank, these officers came to fill most of the senior supply and transport appointments in home and overseas commands. Their experience and guidance were invaluable in the training of young officers. Many were regular contributors to the *Royal Army Service Corps Quarterly*, which published their war-time experiences and their views on the future development and organisation of the Corps. Colonel (later Major General) W. K. Tarver who, for a few months in 1919, had temporarily occupied the seat of the Director of Supplies and Transport pending the return of General Carter from France, and later, in 1925, became Inspector and one of the Colonels Commandant, Colonel (later Major General) Sir Evan Gibb, Brigadier General W. S. Swabey, Colonel W. Elliott, Colonel O. Striedinger, Colonel Sir Harold Percival, Colonel W. M. Parker, Colonel H. G. Burrard, Colonel F. G. Reid and Colonel (later Major General) M. S. Brander all played their part in maintaining the prestige of the Corps and in influencing its development.

Major General Sir Herbert N. Bunbury KCB, the first ASC Major General.

Lieutenant General Sir Frederick T. Clayton KCB KCMG, Colonel Commandant RASC, 1919-1925.

Major General Sir Evan E. Carter KCMG CB MVO, Director of Supplies & Transport, 1919-1921.

Colonel Sir Edward W. D. Ward Bart GBE KCB KCVO, Chairman, Royal Army Service Corps Club, 1919-1928.

Major General G. F. Davies CB CMG CBE, Director of Supplies & Transport, 1925-1929.

Major General P. O. Hazelton CB CMG, Director of Supplies & Transport, 1921-1925.

A Persian fakir weighing up British troops in 1923.

Alarms and Excursions III

The official end of the war with Germany, formally declared by an Order in Council on 10 January 1920, the date when the Treaty of Versailles was ratified, did not immediately signify the end of hostilities everywhere.

The third Afghan War began in May 1919 and, although it did not last long, was the prelude to a troublesome two years in Waziristan.

Total defeat of the Turks in Syria and Mesopotamia did not lead to an immediate pacification of the whole country. Unrest persisted, particularly in Mesopotamia and on the Persian lines of communication. The equivalent of three divisions from India had been left to keep the peace, but these proved insufficient when sporadic disturbances, reflecting dissatisfaction among the Arab populations, in the summer of 1920 became open rebellion. Reinforcements were hurriedly sent from India and a potentially dangerous situation was averted. The disarming of 130,000 tribesmen was a lengthy task and placed heavy demands on the Royal Army Service Corps and the Indian Supply and Transport Corps, as the success of the operations depended on mobility and surprise and extended over a wide area. Units were short of personnel, owing to demobilisation and the lack of replacements. Transport had been reduced to a level required for static conditions, and reserves of supplies had been run down as the size of the force contracted. The rapid arrival of reinforcements from India and a return to mobile operations placed a great strain on the resources at the disposal of the newly-arrived head of the Supply and Transport Directorate, Colonel H. G. Burrard. That conditions could be restored to something like normal within a few months was to a great extent due to the energy and resourcefulness shown by the Supply and Transport Directorate and the commanders of RASC units.

Sporadic outbreaks continued, particularly among the Kurdish tribes, over the next few years, and the Kurdish operations in 1924 were on a scale which merited the award of a clasp to the General Service Medal. Mesopotamia, or Iraq as the country was officially renamed in November 1921, when Feisal became its first king, was entrusted to Great Britain as a mandated territory in 1919. The administration was placed under the Colonial Office and its defence put largely in the hands of the Royal Air Force. 1115 MT Company continued to serve in Iraq until August 1923, when it was disbanded, and a detachment of 12 Fiat lorries (later replaced by Ford 1-ton trucks) was attached to No 5 Armoured Car Company RAF.

In 1926 a preliminary treaty was negotiated between Great Britain and Iraq, and in 1930 an Anglo-Iraq Treaty, to last 25 years, was signed. Under this treaty of mutual protection, Great Britain was granted the use of two air bases, one in the neighbourhood of Basra, and the other west of Baghdad at Habbaniyah. Officers of the Iraq Army and Air Force were to receive instruction in British military schools and establishments, and Great Britain would provide Iraq with Naval, Army and Air Force officers in an advisory capacity.

Conditions in the country so far improved that on 3 October 1932 the British mandate came to an end and on the following day Iraq joined the League of Nations.

Among the few Royal Army Service Corps officers who were seconded to the Colonial Office for service in Iraq were Captains Goodfellow and Eassie, both of whom were restored to the Establishment when the mandate ended. The appointment on the British Military Mission of Inspector of Mechanical Transport, Iraqi Army, was held by Major G. A. Sims from 1934 to 1937, and from 1937 by Captain J. H. Jeffers.

King Feisal died while he was abroad on 8 September 1933 and his body was brought by the Royal Navy to Haifa, where the detachment of 14 Company RASC was given the unusual task of providing an improvised hearse, to convey him from the port to the airfield.

The opening of the oil pipe-line from Iraq to Haifa in January 1935 provided a new and attractive target for the rebellious Arabs in Palestine and one more headache for the forces of law and order.

Up to the outbreak of the Second World War Iraq was one of the more stable territories in the Middle East.

In Ireland there was no peace. In 1914 the Home Rule Act had been put into suspension until the end of hostilities, on the assumption that the war would not last long. By 1916 the Sinn Fein Party lost patience and broke into open rebellion at the Easter Rising. Although the revolt was ruthlessly suppressed, the Republicans were unwilling to make any concessions; when the Government of Ireland Act 1920 was passed in place of the Home Rule Act, the greater part of Ulster (except for the counties of Donegal, Monaghan and Cavan) electing to remain united with Great Britain, the Sinn Fein Party rejected it outright and a state of war between Ireland and England existed until a peace treaty was signed on 6 December 1921. Arthur Griffith became President of the Irish Free State in Southern Ireland, but he died in 1922 and was succeeded by William Cosgrave, who was President until 1931. Finally the Irish Free State Agreement Act 1922 enforced the treaty of 1921, conferring Dominion status on Ireland.

Throughout, service in Ireland was uncomfortable and trying, particularly for the Royal Army Service Corps, whose units had to cover enormous distances and whose vehicles were constantly being ambushed and presented easy targets for sudden attack. The Corps suffered its full share of casualties, many of them fatal. Barbed wire, armed convoys, interrupted communications and a strict curfew were the lot of the soldier in Ireland. But, in spite of the dangers and the discomforts, Royal Army Service Corps units remained astonishingly cheerful and made their own entertainments in barracks, maintaining a full programme of sport and recreation; the keen horseman or fisherman could still find plenty of opportunities to exercise his hobby. Even the 'wind-swept, rain-sodden, sheep-infested, brass-hatted, dreary waste', as one soldier described the Curragh, could provide hunting with the Kildare Harriers, and flat races were well supported.

GHQ Ireland was located in Dublin and, apart from Dublin District, there were a Northern District Headquarters at Belfast and a Southern District Headquarters at Cork. The 1st Division was deployed in the Belfast District, the 5th Division at the Curragh and the 6th Division in the Cork District. Brigade Areas and Sub-Districts were established at places like Londonderry, Limerick, Athlone and Fermoy (which later became a District, including Waterford Area and Clonmel Sub-District). Athlone Sub-District, under the command of the 5th Division, covered the entire Province of Connaught and the Counties of Westmeath and Longford, and 1169 MT Company carried out details over nearly 10,000 square miles.

Incidents were of almost daily occurrence. On 20 July 1920 rioting between Sinn Fein and Union Party workers broke out in Harland and Wolff's shipyards and rapidly spread to the Falls Road district of Belfast, causing £250,000 of damage. It was reported that 'football teams proceed as if they were on active service'. It was a common saying that there were only three ways to get away from Ireland: by discharge, by court martial or in a coffin.

Among those serving then in Ireland were Colonel W. S. Swabey DDS and T, Irish Command, Lieutenant Colonel (later Brigadier) C. Burton, AD of S and T, Lieutenant Colonel H. W. P. Stokes, and Lieutenant Colonel R. J. Slaughter, (both OC RASC Fermoy District) and Major L. C. Bearne, OC 1155 MT Company, Cork, who had been awarded the Albert Medal for gallantry in Serbia in 1916.

When the peace treaty was signed, 'the settlement of the Irish Question' was optimistically announced, to be closely followed by the evacuation of the British Army from Southern Ireland. Already in February 1922 units in and around Cork were busy making plans to leave. In March 26 and 28 (HT) Companies left Fermoy for final disbandment and only 1170 MT Company and a small detachment of 'D' Supply Company were left. A month later the Horse Transport Companies in Dublin were disbanded, and by mid-May the Curragh was finally evacuated by British Forces. All RASC companies were disbanded and the officers and men distributed to units in Ulster or elsewhere.

A fitting epitaph to this phase in the history of the Royal Army Service Corps was given by Lieutenant General Sir E. P. Strickland KCB CMG, GOC 6th Division, at his farewell inspection of 1155 MT Company at Cork on 1 March 1922:

'This Division, as you know, is in process of being broken up, and I hope that we shall all be able to enjoy a more peaceful life. I should like to say that this war has been without precedent and without any of the glory of war. You have been called upon to perform duties repugnant to British Soldiers. Most of you were in the Great War in France and elsewhere and there you faced your enemy and knew what to expect; you were in danger — I might say great danger — and you were prepared for it. The British soldier fights clean, but here you never knew what was going to happen; the fellow who asked you out to dinner would shoot you in the back as you were going out. In spite of it all, you have maintained the traditions of the British Army, discipline and *esprit de corps*.

'I know all of you have carried out your duties in spite of the danger; and I know of several instances where you have been out, and in the early stages were unarmed, yet I recall one, when the postal Crossley was ambushed and several of the escort killed, yet the driver was first out, got hold of a rifle and ammunition and beat off the assailants until assistance came. It is such men as you make the British Army and keep up its traditions.

'I wish to thank all officers, NCOs and men, for the way they have carried out their duties. I have taken a great personal interest in your own work and have watched you and your doings and, further, I was glad to see that you had been able to get in your recreation, and it was with great pleasure that I have witnessed your prowess in the football field. I have been to many regimental sports but I can say that yours were the best I have ever seen.

'In conclusion, I will say that every one of you has done his duty — more I cannot say of any soldier.'

At about the same time another chapter in the history of the British Army was coming to an end. In June 1921 the Imperial Conference, meeting in London, decided that the South African Military Command should cease to exist on 1 December, from which date the Union Government of South Africa had undertaken to be responsible for its defence. It was 126 years since British troops first landed in South Africa on 16 September 1795.

From 1803, occupation by the British Army had been continuous. Now, on 23 December 1921 the GOC and his staff said a last farewell and embarked on the RMS *Arundel Castle*. This was the end of service in South Africa for the Royal Army Service Corps.

After most of the Allied troops had left the Balkans, it was still necessary to keep an Allied force in Constantinople, until the peace treaty with Turkey was ratified. The Treaty of Sèvres between the Allies and Turkey was signed on the 10 August 1920, but it was never ratified. Mustafa Kemal Ataturk, who had had a brilliant war record and had played an important part in the defeat of the Allied landings on the Gallipoli Peninsula, set up a Nationalist Government in Ankara in October 1920 and felt strong enough to repudiate the terms of the treaty in November 1920.

The Greek offensive in Asia Minor in March 1921 fully occupied the Turks until the summer of 1922, when their counter-offensive west of Ankara ended in the complete defeat of the Greeks and the armistice signed at Mudania on the Sea of Marmora in October. Flushed with success, the Turks threatened to occupy the Gallipoli Peninsula.

The British Corps of Occupation, Constantinople, under the command of Major General T. O. Marden CB CMG, did not have the resources to deal with this threat. The RASC component consisted only of 780 MT Company, a HT Company and 'E' Supply Company.

A force was hurriedly assembled and sent to Chanak, on the eastern shore of the Dardanelles, the first contingent coming from Malta, where several transports *en route* to the east were diverted, and practically the entire fighting strength of the garrison of the island was embarked. One member of the small RASC detachment which went with the force from Malta was Lieutenant C. O. Henfrey. 9 (HT) Company, which had been disbanded in 1922, was re-formed at Aldershot as 9 MT Company, equipped with 30-cwt lorries, fitted with the latest pneumatic tyres, and despatched to Constantinople as part of the force, remaining there for most of 1923. Later it was joined by 8 MT Company, which had also been re-formed at Aldershot in 1922.

The Turks were not prepared for further trouble and eventually, on 24 July 1923, the Treaty of Lausanne was signed, and ratified in the autumn of the same year. By the end of 1923 most British troops had left Turkey.

ABOVE: General view of British Army Depôt at Kelia Bay, Turkey, in 1922. OPPOSITE ABOVE: British troops viewing the ruins of ancient Babylon in 1923. BELOW: Khora Creek, Basra, in 1926.

ABOVE: The British Residency, Baghdad, in 1930. LEFT: Major General H. C. Goodfellow CB CBE. In 1930, in the rank of Captain; he was seconded to the Colonial Office for service in Iraq and later became Director of Supplies & Transport, 1951-1954. RIGHT: The Albert Medal.

Cavalry Barracks, The Curragh. (MHY)

Eastern Approaches IV

Under the Irish Peace Treaty of 1921 Great Britain was granted the right to occupy certain off-shore defences, in order to control, in time of war, the Eastern approaches to the Atlantic. These consisted of Spike Island and the outlying forts at the mouth of Cork Harbour, and Bere Island, in Bantry Bay, off the port of Berehaven. In addition to these two Southern Irish ports, the base in Lough Swilly in County Donegal was also included, and Forts Lenan and Dunree were occupied by British Army detachments. Communication with these two outlying ports in Irish Free State territorial waters was maintained from Londonderry by means of a Crossley tender.

The ports of Queenstown and Berehaven were refuelling bases from which Royal Navy destroyer flotillas could extend their range westward into the Atlantic. Lough Swilly provided similar protection for the approaches to the Clyde and Mersey.

Headquarters of the South Irish Coast Defences was on Spike Island, which contained Fort Westmoreland, and is about 100 acres in extent and a mile in diameter. At the southern end was an old convict cemetery. The outlying defences were Forts Camden, Carlisle and Templebreedy.

The Irish name for the island is 'Inispic' — meaning 'the island of the sharp point', although it was more often called the 'Isle of the Saints'. Legend has it that St Mochuda, Bishop of Rahan and Lismore, built the ancient church there. The British Government bought Spike Island in 1779 and a battery of twenty-one 24-pounder guns was erected to command the mouth of Cork Harbour. Fort Westmoreland was named after the Earl who visited the island in 1790, when he was Lord Lieutenant. In 1847 it became a convict settlement and, during the troubles, before the signing of the Peace Treaty, it was used as an internment camp for Sinn Feiners.

The only way to reach the mainland is by boat to Queenstown (Cobh), the nearest town, and Forts Camden and Carlisle at the harbour entrance. Cork is approximately one hour away.

In addition to the Royal Garrison Artillery, with its four coast batteries, there was a Fortress Company, Royal Engineers, at Fort Camden, and detachments of R Signals, RAMC, RASC, and RAOC.

The Royal Army Service Corps detachment consisted of a captain (OC RASC, South Irish Coast Defences), a subaltern (Transport Officer) and six or seven other ranks, with a civilian barracks staff. In addition there were the civilian officers and crews of the War Department Fleet, which consisted of two steam launches, two motor boats, one motor barge and a hospital boat.

Although the island is so small, there were good recreational facilities — a concrete cricket pitch, tennis courts and a nine-hole golf course — and fishing and sailing were always available. Queenstown could offer a variety of amusements. There was no reason why life should be dull on Spike Island or why the garrison should suffer from claustrophobia.

Bere Island (or Bear Island), an out-station of Spike Island, is at the western end of Bantry Bay, some thirty miles from Bantry. The island is seven miles long and about two miles wide at its broadest point. It is hilly and rocky and only sparsely cultivated. The civilian population in the 1930s amounted to about 1,000, some of whom were employed by the garrison as carters and by Royal Engineers on engineer services. Castletownbere is a small place on the mainland about half an hour's run by boat from Laurence Cove, where the War Department pier, transit sheds and RASC barracks store were situated.

The garrison consisted of two batteries of Royal Garrison Artillery, a detachment of Fortress Company Royal Engineers and small detachments of the Royal Corps of Signals, Royal Army Medical Corps and Royal Army Service Corps, all of them administered by the Royal Artillery.

The garrison was in wireless communication with the Headquarters South Irish Coast Defences on Spike Island, which was about 100 miles away in Queenstown Harbour. A weekly Royal Navy destroyer service kept the garrison in touch with Headquarters.

The Royal Army Service Corps detachment consisted of a subaltern, a sergeant and four private soldiers, together with the crews of the War Department fleet vessels and the civilian barracks staff. In 1928 there was great excitement when the old War Department vessel *Cambridge* was replaced by a new very superior craft, the *General McHardy*. In addition, Royal Army Service Corps transport consisted of the War Department launch *Magpie*, two GS wagons, a mess cart and two light draught horses, Boxer and Tall Boy. In 1934 there were plans to replace the horse transport by a six-wheeler lorry.

The somewhat isolated garrison of Bere Island was, perforce, rather self-contained for its amusements. Shooting, fishing and sailing were always available and the usual recreational facilities were at hand. Twice-weekly trips were arranged to Castletownbere. But the social life did not come up to the standard of Spike Island, where garrison dances were regularly held, with young ladies imported from Queenstown.

Service in the South Irish Coast Defences presented many paradoxes. Although the area was outside the United Kingdom, it was treated as a 'Home Station'. The garrison belonged to the land forces, but its members spent most of their time 'at sea'. Geographically the islands were situated in South-West Ireland, but militarily and officially they were within Western Command of the United Kingdom.

This life of fantasy did not last long. Sixteen years after the signing of the Anglo-Irish Peace Treaty, talks began early in 1938 between the British Government and Mr de Valera, and an agreement was signed on 25 April, under which Great Britain surrendered her rights to occupy the Southern Irish ports and the base in Lough Swilly. Only Churchill stood up and violently opposed this 'gratuitous surrender of our right to use the Irish ports in war', which he regarded as 'a major injury to British national life and security. A more feckless act can hardly be imagined — and at such a time'.

Not many members of the Royal Army Service Corps had the opportunity of serving in the South Irish Coast Defences — the officers can be counted on three hands — but those who did can look back on three years of service in one of the more unusual of Royal Army Service Corps stations.

Brigadier General Lord Montagu of Beaulieu, Adviser on Mechanical Transport to the Government of India, 1915-19. (RE)

The North-west Frontier V

Before 1916, members of the Army Service Corps did not serve in India. The Supply and Transport Corps of the Indian Army, formed in 1902, was responsible for carrying out the equivalent functions, although there were major differences in the constitution and status of the two Corps. In contrast to the Army Service Corps, which had for many years been recognised as a combatant corps and had distinguished itself during the South African War, the Supply and Transport Corps in 1914 was still a departmental branch, and not geared to the support of large-scale operations overseas.

The responsibilities of the Supply and Transport Corps were wider than those of the Army Service Corps and included the handling of stores, which in the British Army were the province of the Army Ordnance Department; and there was a great deal of accounting and auditing for which the Army Service Corps was not responsible. Those stores included, in addition to reserves of food and forage, the mobilisation clothing of British and native units and many items of mobilisation equipment, such as harness, saddlery, transport vehicles and ambulances, blankets, sheets, hospital stores and fuels.

The executive branch of the organisation consisted, in 1914, of 400 officers, a number of British warrant and non-commissioned officers and a large number of Indian clerks, storekeepers and agents, and the native establishment of the various transport units. The officers were found from transfers from Indian regiments or British units serving in India, on a five years probation period. The inducement was a higher rate of pay, but the number and quality of permanent transfers did not measure up to the requirements of a Corps that was to take part in a major war. A number of Army Service Corps officers were seconded from time to time for five years, but they only maintained a tenuous connection between the two Corps.

The Quartermaster General in India was responsible for the Supply and Transport Corps, with a Director of Supplies and Transport at Army Headquarters, who was a major general. Beneath him were representatives at divisional headquarters, whose functions corresponded to those of Assistant Directors of Supply and Transport of Commands in the Army Service Corps. Each division had a Divisional Supply and Transport Officer, who was a colonel or lieutenant colonel (in some cases there were two officers, one for supply and one for transport), and who commanded all the personnel of the Supply and Transport Corps in the Division. He had to assist him a divisional store officer, who usually acted as his adjutant, a divisional contract officer and a station supply officer.

At lower formations, brigades and outstations, there was a Brigade Supply and Transport Officer, who combined all the functions of the Divisional Supply and Transport staff, with the assistance of a number of warrant officers and non-commissioned officers. His was a busy life, as he was responsible for the major part of the administrative services in his station, including all the supply and transport arrangements for manoeuvres in the brigade area.

The supply service made considerable use of civilian contractors, a system which, with strict control, may have met the needs of peace-time conditions, but left no opportunity for the supply officer to practise in peace his war-time duties of requisitioning and distributing supplies. It was also the custom for regimental officers to appoint their own contractors for dealing with their supply and transport requirements, even when a regiment was involved in operations on the North-West Frontier.

This practice continued in India until after the Great War, when a system of supply depôt companies and divisional supply companies was introduced, which could meet both peace and wartime requirements.

In India the quantity of transport which each State was capable of providing was well known. Special supply and transport officers had the responsibility of enumerating the available mules and camels throughout the country. All that the transport officer needed to do was to give notice of his requirements to the Cantonment Magistrate or the Resident Commissioner, who was then responsible for collecting and delivering the required amount of transport at the appointed time and place.

Thus the work of the supply and transport officer was, more often than not, that of a co-ordinator, an agent or an accountant, rather than that of a military officer in the field.

Each transport corps (or unit) was commanded by an officer (usually a major), who was available to assist the formation S and T staff in transport matters generally. Most of the transport units were only in a cadre state, capable of expansion if required, although on the frontiers and in up-country Divisions, units were maintained fully mobilised at all times.

Forms of transport varied from mule carts and pack mules to camels and elephants, but before 1914 there was virtually no mechanical transport in the Indian Army. Various trials had taken place during the previous ten years to test the feasibility of using mechanical transport for military operations but, owing to the nature of the country, the unsuitability of existing roads for the heavy vehicles then available, and the almost complete lack of any development for commercial purposes, mechanical transport offered few practical advantages for the foreseeable future.

The Supply and Transport Corps organisation was then more centralised at divisional level than the Army Service Corps, but it suffered by comparison in not having a field organisation fully trained or designed for active military operations other than frontier warfare.

Early in the war the need for mechanical transport became apparent, and some of the lost ground was made up by the energy and initiative of Brigadier General Lord Montagu of Beaulieu (the second Baron), who was appointed Adviser on Mechanical Transport Services to the Government of India in 1915, and remained in the appointment until the end of the war. He had been a great pioneer of motoring in Great Britain and his energetic advocacy had helped the development of the modern motor car. In addition to placing orders direct with manufacturers for lorries from England and Ford trucks from America, he saw at once the vital necessity of establishing adequate workshops in India, and was responsible for their construction at Rawalpindi.

These measures, however far-sighted, were not in time to assist the military situation in Mesopotamia, where the fall of Kut el Amara on 29 April 1916 was largely attributed to a breakdown of administrative support. This led to a complete overhaul of the lines of communication. Colonel Beadon gives a full description of the shortcomings of the supply and transport services and describes the energetic steps taken to put matters right and to place the administrative services on a sound basis. Brigadier General P. C. J. Scott, an Army Service Corps officer, was appointed Director of Supplies and Transport in August 1916, with Colonel W. M. Parker, who had done a tour of duty with the Indian Army before the war, as his deputy. The first Army Service Corps unit had arrived in Mesopotamia in January 1916 from Egypt (596 Company), and this was soon followed by others from England. This was the start of the first real collaboration between the Army Service Corps and the Supply and Transport Corps.

At about the same time it was decided to send Army Service Corps units to India, to make good the shortage of mechanical transport. Four companies, including a motor ambulance convoy, were sent in May 1916. These companies (630 MAC and 692, 693 and 694 Companies ASC) were kept busy during the next year and their vehicles (3-ton Albions and Fiats, Thornycrofts and 1-ton Star trucks) were thoroughly tested under Indian conditions and stood up to them remarkably well, thanks largely to the workshops at Rawalpindi. They operated with the Waziristan Field Force from Dera Ismail Khan and from Bannu up the Tochi valley. The Motor Ambulance Convoy (630 Company) was in continuous operation with its Daimler ambulances during 1917 and 1918.

At first there was no Army Service Corps depôt for the administration of Army Service Corps personnel serving in India. In 1917 one was established at Cornwallis and Pekin Barracks at Bangalore, in Southern India, from 871 Depôt Company which had arrived from England. This depôt took over the administration of all Army Service Corps personnel serving in India, and maintained the strength of Corps units in India from drafts from Bulford MT depôt in England. The Indian Government had formed two mechanical transport schools at Meerut and Rawalpindi, in order to meet the great demand for MT drivers in India, Mesopotamia and Persia. The instructional staff for these schools was provided from the ASC Depôt at Bangalore. Several hundred Army Service Corps NCOs and men were also posted for duty with MT companies formed by the Indian Government. Horse transport NCOs and men were attached to mule, pony and 'ekka' (two-wheeled AT cart) corps (the equivalent of a company) of the Supply and Transport Corps, for duty as supervisors of Indian personnel. The depôt also served as a leave centre for men serving in Mesopotamia and Persia, who spent their month's leave in India.

The commanding officer was Lieutenant Colonel J. H. B. Peyton, who established a close and harmonious relationship with the Supply and Transport Corps and later transferred to the IASC. The depôt published an excellent magazine, the *Indian ASC Monthly*, which appeared from October 1917 to February 1919.

At the end of the war more Army Service Corps units arrived from East Africa and Mesopotamia, to add to the growing number of MT units operating in India.

Operations on the North-West Frontier during 1919 and 1920 were important to the Royal Army Service Corps and the Supply and Transport Corps, for this was the first time mechanical transport was used on a large scale for the maintenance of troops in forward areas.

Most of the RASC units in India were employed on the North-West Frontier, with their base at Chaklala, three miles from Rawalpindi. Very few of the British officers and men of the Corps who served in India missed the experience of service on the frontier.

Owing to the mountainous nature of the terrain and the lack of good communications, it had not been possible to establish a secure administration in the wild country to the west of the Indus Valley. In between the North-West Frontier Province and Afghanistan lay the mountainous tracts of Waziristan, with its ill-defined boundaries and numerous tribal areas, jealously guarded by their well-armed and quarrelsome tribesmen. Theirs was not a peaceful existence and their lives were governed by feuds, pillage and plunder. The Indian Government had for many years accepted that a policy of non-interference was the only practical method of administering these territories, but it was a policy which depended for its success on friendly relations with Afghanistan. The main object was to contain the unruly elements within their own boundaries, and to prevent raids and incursions into British administered territory.

Some protection of the trade routes up the river valleys and through the passes was obtained by government subsidies to the tribes, in exchange for promises of non-interference with traffic, but these did not always overcome the tribesmen's inherent greed and their natural inclination towards violence. And so it was necessary to establish fortified posts and blockhouses at strategic points, and these were garrisoned by the militia of North and South Waziristan, with HQ at Miranshah and Wana; local levies were raised for the express purpose of restraining their fellow tribesmen.

These tribal forces were supported by regular troops, in 1919 based on Peshawar, which dominated the approach to the Khyber Pass into Afghanistan and the Kohat Pass into Waziristan, and on Bannu and Dera Ismail Khan (DIK or 'Dreary Dismal' to the ordinary soldiers). From time to time, when raids by tribesmen became too troublesome, it was necessary to make a show of force, and punitive expeditions went into the tribal fastnesses. The effect of these operations was never more than temporary. The nature of the country favoured the tribesmen and, although many of them suffered the loss of what little property they possessed, most lived to fight another day. It was not until the policy of building better communications and establishing fortified camps deeper inside tribal territory was adopted that a greater measure of stability on the frontier was achieved.

The delicate balance which had been maintained by friendly relations with Afghanistan was rudely shattered in May 1919 when Amanullah Khan allowed his ambitions to carry him away, and attempted to extend his suzerainty over the tribes on his south-eastern frontier.

The Afghans started the war on 6 May 1919 by attacking the western end of the Khyber Pass. For three days the situation depended on the staunch resistance of the Khyber Rifles, the irregular militia, who held the forts dominating the pass until the arrival of regular troops.

A second Afghan army under General Nadir Khan advanced into Khost and attacked the fort of Thal, at the southern end of the Kurram Valley. This attack was beaten off and the army defeated by a relieving force under Brigadier General Dyer, who had been in command at Amritsar.

Requests for an armistice were received on 31 May and three days later the war was over. A peace treaty was signed on 8 August 1919 and a second treaty at Kabul on 22 November, re-affirming the Durand Line as the frontier between India and Afghanistan. This 'frontier', which had been negotiated by Sir Mortimer Durand with the Amir of Afghanistan, Abdul Rahman in 1892, and had been demarcated on the ground for most of its length during the following two years, was really only an administrative boundary; many of the tribes, whose territory lay on the Indian side of the line, were in no way subject to the Government of India.

Although the Third Afghan War was of short duration, there was fierce and bitter fighting while it lasted, and, for a time, the situation looked serious. The loyalty of the tribesmen, on whom peace in the area depended, was badly undermined and the operations had an unsettling effect on all the tribes in the frontier area. The next four years were an active and busy time for the troops of the Waziristan Field Force.

During the Afghan War, mechanical transport companies were employed on the lines of communication from around Peshawar through the Khyber Pass to the troops in Afghanistan. One incident illustrates what was probably the first use of wheeled transport in an operational role on the frontier. On 7 May the Somerset Light Infantry moved forward to reinforce a small column of Sikhs and Gurkhas who, with a mountain battery, had occupied Landi Kotal. They were carried in 'hooded' lorries, which proved to be a modern version of the Trojan Horse, as both the tribesmen and the Khyber Rifles thought that they merely contained routine replacements of stores, and allowed them to pass unmolested. The Afghans let slip a golden opportunity, but no one in the column realised what a risky operation it was.

The important part played by mechanical transport in these operations can be appreciated when one considers that over 400,000 pack animals were required for the support of frontier warfare on this scale. In the Tirah Campaign in 1897/8 a force of 44,000 combatants had required 60,000 animals. Pack transport was extravagant in space and required round-the-clock protection. Its own demands for forage and other stores seriously added to the overall payload. The length of the administrative tail imposed a serious brake on the mobility of the force, and completely destroyed the possibility of surprise.

At this time most of the Royal Army Service Corps companies had 25% Indian drivers (from the Training School at Meerut), the remainder being British. Later these proportions were reversed.

The railway stopped short of the Khyber Pass, which was about 20 miles long to Landi Kotal. The road through the pass, which climbs about 3,000 feet, had a good surface but was narrow in places, which made passing a camel convoy a delicate business.

The lines of communication for the maintenance of the field force from the plains of the Indus Valley into Waziristan followed the main water courses. The chief of these were along the River Tochi and the River Gumal. The trans-Indus narrow-gauge railway ran as far as Bannu along the former and to Tank and Kaur Bridge along the latter. A total daily tonnage of 600 tons could be delivered at these railheads. But forward of these points only Ford vans, in addition to animal transport, could operate, from Bannu up to 40 miles and from Tank for 20 miles. The tonnage delivered by rail was supplemented by the use of a Decauville railway from Dera Ismail Khan to Tank (a distance of 42 miles) and by mechanical transport following the same route.

As a result of the Third Afghan War, and the campaigns in Waziristan which followed it, the policy of containment and non-interference devised by Lord Curzon was abandoned. It was decided that strong forces would be established beyond the passes, at Landi Khana near the Afghan frontier, north-west of the Khyber Pass, and that the railway would be extended from Jamrud through the pass. In order to curb the activities of the Mahsuds, one of the most troublesome of the tribal races, a circular motor road would be built and a permanent cantonment established at Razmak from which the surrounding country could be effectively controlled. This 'roads policy' in Mahsud and Wazir territory was first sanctioned in 1922, when Lord Peel was Secretary of State and Lord Reading was Viceroy

When the operations in Waziristan which immediately followed the Third Afghan War drew to their close in 1921, the importance of mechanical transport as an essential element in the successful prosecution of North-West Frontier operations had been fully demonstrated. It was seen that it was not only necessary to improve road communications to enable mechanical transport to be used further forward, but also to build lateral communications between Tank and Bannu. Out of this necessity grew the concept of the Waziristan circular road, which later became so well known to members of the Royal Army Service Corps in India. This road, when it was completed, provided a direct link between Dera Ismail Khan and Bannu (1,230 feet) and then curved westward in a great bend through Idak, Razani and Razmak and back to Khirgi, Manzai and Tank. The route wound through hilly, undulating country and climbed from 2,384 feet at Idak to over 5,000 feet at Razani, reaching its highest point at Razmak, which is 7,948 feet above sea-level. The next 24 miles dropped some 4,000 feet to Sararogha on the way to Khirgi. While the road was being constructed there was a continuous demand for transport for the carriage of materials, and for the maintenance of the troops engaged in building and protecting it.

During this period the personnel of Royal Army Service Corps MT units became merged with those of the Supply and Transport Corps and the two corps became largely integrated. Units lost their original identity and became Mechanical Transport Units India or Indian (MT) companies, but they continued to be classified as Royal Army Service Corps units. For instance, 22 MT Company (formerly 694 Company), which in March 1921 was at Kohat, on the North-West Frontier, 40 miles from Peshawar, consisted of four British officers, six senior British NCOs, and the remainder of Indian drivers and artificers. The drivers were sepoys — properly 'enlisted' men — but the artificers were 'enrolled' and their NCOs were called 'foremen' (three stripes) and 'charge-hands' (two stripes).

When a start was made on building the road, operations were proceeding against the Mahsuds, and Headquarters of 'B' (MT) Column and 9 MT Company were at Dera Ismail Khan, 7 MT Company was at Manzai and 12 MT Company was at Khirgi, and other MT units at Bannu. Manzai was a typical frontier station and consisted of seven bungalows; one was a hospital, and the troops stationed there were one native infantry regiment, one company South Waziristan Militia, three MT companies (two equipped with Ford vans and one with motor ambulances), the hospital and HQ 10th Infantry Brigade. All the British other ranks lived under canvas and a good many of the British officers too. All the offices were in tents. The post was barbed-wired all round, with sangars every 300 yards. The perimeter was about a mile round, and no-one was allowed outside the wire after 4 pm.

The Fords were used mostly on transporting supplies from railheads, first at Kaur Bridge and afterwards at Khirgi; they also took part in road protection duty, manned with Lewis gun teams. The average turn-round of a van per day was about 40 miles and they worked from 5.30 am until 12.30 pm daily. This was trying, as the shade temperature averaged 115 degrees. They had to parade again at 6.30 pm to get their vans ready for the next day. The routine went on unceasingly, every available van on the road daily, with no holidays of any sort while the force was engaged on operations.

Lieutenant B. U. Farrell of 11 (MT) Company died of cholera during an outbreak in Manzai, Tank and Khirgi in June 1921. Only a year before in 1920 he had won the light-heavy weight title in the Inter-Services Boxing Championships at Olympia.

The contribution which the Royal Army Service Corps made to India was considerable. In November 1920 there were 98 RASC officers serving in India (one colonel, 8 lieutenant colonels, 17 majors, 36 captains and 36 subalterns). By February 1923 the total had reached 130. Officers detailed to embark for service in India during the 1920/21 trooping season totalled 43. Among them was Colonel F. M. Wilson who, on 17 August 1921, was appointed Director of Supplies and Transport, India, and promoted to major general. Also on the list were Brevet Major (later Brigadier) C. L. St J. Tudor, Major (later Lieutenant Colonel) W. R. V. Warren and Captain (later Major General) F. S. Clover.

The appointment of Adviser to the Government of India for mechanical transport services, previously held by Lord Montagu of Beaulieu, was filled by a Royal Army Service Corps officer, Lieutenant Colonel W. P. Robinson, DSO, OBE, who was graded as an Assistant Quartermaster General, until the end of January 1924. Later a similar appointment was made at the India Office, under the title of Mechanical Transport Adviser, and held by Colonel T. M. Hutchinson DSO OBE MIMechE from June 1928 until 1930.

Colonel W. M. Parker returned once more to India in 1925 and joined General Wilson's staff at Army Headquarters, Simla, as Deputy Director of Transport, an appointment which he continued to hold under General ap Rhys Pryce until the Royal Army Service Corps left India in 1928.

Under Major General Wilson the two corps became even more closely integrated and, during his four years as Director, the climate became more and more favourable for an ultimate merger.

At Dehra Dun the Commandant of the Mechanical Transport Training School (which moved to Sitapur in October 1921) was a Supply and Transport Corps officer, the second-in-command and chief instructor a Royal Army Service Corps major. The rest of the training staff included four RASC captains, including W. d'A. Collings (later Major General), three subalterns and fifty WOs, NCOs and men from the RASC. The Indian drivers' course consisted of two months' foot-drill, one month in the chassis class and one month in the driving classes. Trained recruit drivers were posted to MT companies in India or Mesopotamia.

The North-West Frontier could claim a (so far) unique form of military transport in the Khyber Ropeway Company. During the war, almost every form of animal and human transport was used, including the highly successful Reindeer Transport Column, which operated in Lapland and North Russia. By means of the Ropeway Company, its headquarters at Jamrud, near the entrance to the Khyber Pass, practically all the requirements of troops stationed at Shahgai, Alimasjid and Landi Kotal were delivered.

The ropeway was originally designed for the Third Afghan War, but initial difficulties and lack of experience delayed its completion until 1920. Major K. J. McMullen RE was sent out by the War Office in September 1919. He was responsible for the successful construction of the first three sections. In early 1920 responsibility for the completion, maintenance and working of the line was passed to the Mechanical Transport Department in India. The Khyber Ropeway Company was formed and Lieutenant A. F. Lobb, RASC was put in command on 1 July 1920, with the acting rank of major. Sadly, he was taken ill in November 1921 and died less than a year later. He carried out the initial organisation of the ropeway as a transport unit and subsequently its running and maintenance. Majors Boutflower (until March 1924), Furneaux and Ainslie succeeded him.

Jamrud, headquarters of the company, lay 1,600 feet above sea level in the Peshawar Valley and was the furthest point then reached by the North-West Railway.

From Jamrud it is three miles over gently rising ground to the entrance of the narrow valley of the Khyber stream, which flows from the pass itself. The route then rises steeply and goes through Shahgai, Alimasjid, Zintar, Nikki Khel and Sultan Khel for nearly 20 miles until it reaches the summit of the pass at the fort and defended camp at Landi Kotal, 3,700 feet above sea level. From Landi Kotal the route descends steeply for three miles to Landi Khana, within about a mile of the Afghan border.

After the Third Afghan War a good motor road was built through the pass. The ropeway was constructed in six sections. Four of the stations were military camps inside perimeter wire — Jamrud, Ali Masjid, Landi Kotal and Landi Khana. The other three intermediate stations were in the open, and stone forts were built to contain and protect them.

Each section of the line had its own moving 'monocable' — an endless steel rope — from which the carriers, each with a load of 3 cwt, were suspended from clips and moved along the line with the endless cable.

The line was supported on standards varying in height from 10 to 150 feet, depending on the terrain, at intervals of between 200 and 800 yards. The cable passed over grooved wheels, mounted in tandem on the two arms of the standards, one for the outward and one for the homeward line. At each station there were arrangements for the carriers to be transferred by a stretch of monorail from one section to the next. The carriers travelled at 35-second intervals. The power was provided by 40hp stationary Crossley engines, which were started on petrol but then ran on kerosene. In each engine-house two or four engines were bedded down and used in turn, to ensure the least possible interruption to the system. There were four power-houses: at station 2 for the first section, at station 3 for the second and third sections, at station 5 for the fourth section, and station 6 for the fifth and sixth sections.

The rope moved at about four miles per hour and the journey from Jamrud to Landi Kotal took about 4½ hours and another half hour to Landi Khana, although the final section was not so busy, there being only a small post at Landi Khana. It was impossible to protect the line at night, so working was restricted to daylight.

The system was not used for passengers, although it would have been most suitable for the evacuation of casualties. Nor were rifles or ammunition carried, for fear of tempting tribesmen. The company was provided with a small transport detachment of six three-ton lorries, five Crossley tenders, and three motor-cycles and a well equipped maintenance workshop. Repair technicians were highly skilled and the minimum of delays in operation were allowed. Once the cable was cut through by tribesmen, with considerable damage to a 70-foot-high standard and the tension trolley and winch at one of the stations; the damage was repaired within 48 hours.

The establishment of the company was five British officers and four warrant officers, with one Indian overseer, four Indian WOs and NCOs, 20 MT drivers, 16 stationary engine drivers, 46 artificers and 456 others.

When the Khyber Railway was completed as far as Landi Khana, the ropeway had served its purpose for six years. Although in Italy and other mountain areas in Europe cable railways, using a fixed rope with a light tow rope and winching apparatus, have been used to support military operations, this was the only case in which a moving monocable has been used to deliver military stores over a distance equivalent to the width of the English Channel.

By the middle of 1923 mechanical transport, in the shape of Royal Army Service Corps units, mostly with 90% Indian personnel, were firmly established in India. They were grouped into a number of MT columns — 'A' MT Column at Peshawar, 'B' MT Column at Dera Ismail Khan, 'C' MT Column at Quetta, 'D' MT Column at Chaklala (three miles from Rawalpindi — 'the Aldershot of India') and 'E' MT Column at Poona. The Heavy Repair Shops (formerly the Base MT Workshops) and Central MT Stores Depôt and Reserve Vehicle Park were at Chaklala. In addition there were three technical inspection teams operating on three circuits: the northern circuit, based on Peshawar, the central circuit, based on Rawalpindi, and the southern circuit, based on Poona.

The mechanical transport companies were organised as follows: ten heavy motor transport companies, equipped with 3-ton or 30-cwt lorries; three light motor transport companies, with Ford vans or staff cars; eight motor ambulance convoys; four medium artillery companies; four mobile repair units. (At Appendix D is a location statement showing the distribution of Royal Army Service Corps units in India at the end of 1922.)

Pressure for the development of mechanical transport in the Indian Army came almost entirely from the need to support operations on the North-West Frontier. During 1923, when the circular road was still under construction, the bulk of the mechanical transport was deployed along the frontier: eight companies on the Dera Ismail Khan line and six on the Bannu line (leaving only one heavy company, three medium artillery companies, three mobile repair units, one light company and five motor ambulance convoys in India proper).

The operations in Waziristan between 1921 and 1924, for which the India General Service Medal was awarded, with the clasp 'Waziristan 1921-24', made continuous demands on Royal Army Service Corps units. The medal was awarded for 'Service in North and South Waziristan, Bannu, Dera Ismail Khan civil districts and that portion of the Mianwali district which lies west of the River Indus, and also the military posts of Mari Indus and Darya Khan, east of the River Indus, between the 21st December 1921 and the 31st March 1924.'

The Waziristan Field Force operated in the country beyond Khirgi (1,700 feet) at the foot of the Suleiman Mountains, in the valley of the Tank Zam. The base was the railhead at Dera Ismail Khan (67 miles away from Khirgi by road). The MT units were divided between Dera Ismail Khan, Manzai and Khirgi. 'B' MT Column Headquarters (Lieutenant Colonel A. G. Galloway DSO), 18 Mobile Repair Unit and 27 (Indian) MT Company (MAC) were at Dera Ismail Khan; 8 (Indian) MT Company (Ford vans) was at Manzai, and 2, 21, 23, 24 and 25 Heavy (Indian) MT Companies were at Khirgi. Large MT convoys operated daily to Jandola,

Kotkai and Sararogha (over 5,000 feet in height and 27 miles from Khirgi). To support the Razmak Field Force in early 1923 MT companies were assembled from all over India, and came under command of 'A' MT Column from Peshawar. 1 Indian (MT) Company and 7 Indian (MT) Company (125 Ford vans) came from Lahore. In addition 14, 16 and 22 Indian (MT) Companies joined the column. The column had a small office with Force Headquarters, with 7 (MT) Company at Tal and 16 (MT) Company at Idak.

As the new road was built Ford vans first began to replace pack transport and, when the surface had been metalled, lorries could operate further forward. Heavy convoys travelled daily from Bannu to Idak and later as far as Razani. Ford vans were working close to areas which had never been surveyed. Large quantities of stores were being dumped daily at Razmak, and there was never enough transport to meet all the demands. Bannu was maintained as a base for heavy repairs, with special artificers and a tyre press.

Service in India could be divided into three distinct categories: first, the regular Indian station, where good barrack accommodation existed, with ample opportunites for sport and recreation and, in some cases, hill stations where the hot weather of the plains could be avoided. Rawalpindi, with the hill station of Murree, came into this category. The second class included frontier service in stations like Quetta and Peshawar, where life was peaceful, and where living conditions were pleasant, with good facilities for entertainment and sport.

Thirdly, there were the frontier posts in Waziristan, where accommodation was primitive — mostly tented — and where the need for security was constant, with barbed-wire perimeters and forts, escorts and pickets; long hours of work, poor roads, frequent flooding and, in the hot months, temperatures rising to 130 degrees in the shade (recorded in Dera Ismail Khan), with limited recreational facilities.

The following contemporary account by a member of 4 Indian (MT) Company RASC, fresh from the fleshpots of Karachi, of life at Idak, a frontier post 27 miles beyond the railhead at Bannu, gives a good idea of conditions on the North-West Frontier in January 1923.

'Imagine a slice of Salisbury Plain, only brown instead of green, about half a mile square, enclose the whole by a mud wall, and a good deal of barbed wire, place in it a mud fort, wireless aerials, many tents and mud bungalows, etc, and an assortment of light cars and lorries, and set the whole amid brown, bare, uninteresting hills, anything from 500 to 2,000 feet high, stretching to the horizon on all sides, and you have a slight impression of Idak. From my office I can see 14 "posts", small square forts, each of which shelters a picquet of 22 Indian troops, who do a tour of duty lasting 14 days, their food and water being taken to them by mule convoys. Along the road and tracks are endless convoys of camels and mules, the former predominating, carrying supplies to the troops at the front and located beyond the effective range of MT, which range is entirely controlled by the existence or non-existence of roads.

'What do we do all day? This can be answered in one word — work. Up here we have 86 lorries, Peerless, Albion and Thornycroft, in addition to 23 Ford ambulances and various cars. The average monthly mileage of each lorry is over 1,000 miles. Our daily runs are to Bannu (27 miles), Asad Khel (28 miles), Dardoni (14 miles), with convoys of 48 to 66 lorries loaded with supplies. The roads are not at all good and very hilly; for instance, from Thal to Asad Khel, a distance of 21 miles, the road rises 5,000 feet. The convoys move off at 7.30 am and by the time they have done their journey out and home, re-loaded at the Supply Depôt and parked up for the night, it is 6 pm at the earliest. Other vehicles are allotted to various standing details. The average number of vehicles in shops is surprisingly small, especially taking into consideration the condition of the road and the age of the vehicles. Our own Peerlesses are now seven years old, and the Albions of 16 (MT) Company are equally venerable.

'During this month, ie, 5th to 28th, we only remained on the park on five occasions, each time the cause being the same, ie, parts of the road had been washed away by rain and it was, therefore, impassable for any heavy transport. On these days vast hordes of 'tin lizzies' functioned in our stead. Our day's work being over, and we having dined, we retire to our various tents and talk or read in bed, covered by as many blankets, overcoats or coats, warm, MS, as we can obtain. At night it is very cold and not too warm during the day.'

Of course, no families were permitted in these frontier stations until the official end of active service conditions in 1924. Dera Ismail Khan, which was the Headquarters of the Wazir Force, was the nearest place where families were allowed (75 miles away), and accommodation there was limited. Six weeks leave to the hills was the normal allowance for British service units, but pressure of work seldom made this possible for MT units. Similarly, MT units could not be given regular reliefs, as was normally the case for other British and Indian units. There were men who had hardly been outside the wire for more than two years.

Conditions in Waziristan improved greatly when the circular road was completed and the building of the iron and stone bridge over the Tochi River, near Tal, was finished by the Royal Engineers in 1924. Once the road was made and a permanent garrison 10,000 strong, consisting of a brigade of six battalions of infantry with artillery support, had been established at Razmak, the situation became more settled and the need for transport considerably reduced. In the opening chapter of *Bugles and a Tiger*, John Masters, who served in Razmak when he first went to India, gives a good picture of life in this frontier station.

By 1926 the Royal Army Service Corps units in Waziristan had been reduced to a Column Headquarters and five companies, located as follows: Dera Ismail Khan: 5 Indian (MT) Company (MAC), with detachment at Manzai; 22 Indian (MT) Company (MRU). Bannu: HQ MT Column; 13 Indian (MT) Company. Idak: 18 Indian (MT) Company. Razani: 17 Indian (MT) Company.

Maintenance of the troops in Waziristan was carried out by regular convoys from Bannu, found from 13, 17 and 18 Indian (MT) Companies which, between them, had seven sections of light lorries (Karriers, Albions and Crossleys) and one section of Ford vans.

The force in Razani Camp totalled about 2,000 All Arms, including an infantry battalion. 17 Indian (MT) Company was commanded by Lieutenant Colonel E. H. Fitzherbert who, by virtue of his Army seniority, was the senior combatant officer in this large barbed-wire camp. He was, therefore, the commander of the whole force. Similarly, his successor, when he was posted back to Chaklala, Lieutenant Colonel J. L. Weston, also became the force commander, as the senior combatant officer in the station.

There is no doubt that service in India was not particularly popular among members of the Royal Army Service Corps. For one thing, five years away from home so soon after the end of the war was not much of an attraction. On the other hand, those who were fond of an open air life and outdoor sports extolled the pleasures of trekking in Kashmir, of hunting in the Vale of Peshawar, and 'poodle-faking' in the hills. Excellent opportunities for shooting, fishing and playing games like polo, golf and tennis existed in most stations, but there was always the bogy of the prospect of two years' service on the North-West Frontier. For the unmarried officer with a small private income there were, no doubt, many attractions but, for the married officer and most of the men it was a trying experience, with the added expense of maintaining two households during periods of family separation, aggravated by the withdrawal of special concessions and allowances when active service was officially terminated on the frontier in 1924.

Members of the Corps, when they were not serving on the frontier, played a leading part in every form of Army sport in India. At one time there was hardly a game or sport in Peshawar in which the RASC was not represented on the administering body. Lieutenant

Colonel C. H. Furneaux DSO was appointed Master of the Vale of Peshawar Hunt; the names of Fitzherbert, Marrison, Tudor, Clover, Warren and others were familiar on the cricket field, the golf course and the tennis court.

Lieutenant General Sir George MacMunn KCB KCSI DSO who, as Inspector General of Communications in Mesopotamia, had worked closely with the Army Service Corps and was now Quartermaster General in India (from 1920 to 1924), was anxious to change the name of the Supply and Transport Corps so that it would conform to the status and prestige of the Royal Army Service Corps, which he had learnt to appreciate.

In July 1923 the Supply and Transport Corps became the Indian Army Service Corps, and speculation was rife that an amalgamation was imminent.

The Government of India had put forward proposals for the reorganisation of the Supply and Transport Corps in 1922. These were designed to change its status from that of a department, to bring it in line with the RASC as a combatant corps. This would make a merger possible. The proposal envisaged an Indian Army Service Corps, in which all the British officers and other ranks would be provided by the Royal Army Service Corps, and the Indian elements found, initially, by transferring the serving personnel of the Supply and Transport Corps. British officers and other ranks of the S and T Corps would have the option of transferring to the RASC. The effect of these proposals would have been to incorporate in the Indian Army, in place of the Supply and Transport Corps, a transport organisation of basically the same pattern as that of the Indian MT Companies RASC, which had already been formed, with the addition of a supplies branch analogous to that of the RASC and manned in the same way. The responsibility for the transport and supplies services of the Indian Army would virtually have been transferred to the Royal Army Service Corps. When the change of title to Indian Army Service Corps was made, the Indian Government's proposals were still under discussion, and the negotiations continued until 1927, before a decision was made.

Why the original proposals were rejected need not concern us, although it was obvious that vested interests were involved. But when the reorganised Indian Army Service Corps was formed by the Indian Government in 1928, it was entirely dependent on the permanent transfer of a substantial number of RASC officers and other ranks. Once the decision had been taken no time was lost, and by the end of the 1927/28 trooping season most of the transfers to the Indian Army Service Corps had been completed.

This solution brought to an end the opportunity for RASC officers and men to serve in India, unless they were prepared to spend their whole service with the Indian Army, and denied to the Royal Army Service Corps the advantage of experience on the North-West Frontier.

In the end, one hundred and thirteen officers, a hundred and two warrant officers, and about seven hundred NCOs and men voluntarily transferred to the Indian Army Service Corps. This represented nearly 20% of the strength of the Royal Army Service Corps, which amounted to a considerable loss of experience and ability to a corps which had already suffered heavily in the reduction of its establishment.

There were many who deplored this solution and who would have preferred the original proposals which would, undoubtedly, have enlarged the scope of the Corps. But, with hindsight, and the knowledge that the British Army in India was to come to an end in 1947, perhaps it was as well that the 'little Englander' alternative was chosen when it was. Whatever the merits of the alternative may have been, the Royal Army Service Corps can be proud of its decade of service in India and the part it played in establishing mechanical transport in that country, and thereby helping to secure more settled conditions in Waziristan and among the tribesmen of the North-West Frontier.

The fact that six of the Royal Army Service Corps officers serving in Chaklala in 1927 later rose to the rank of major general is sufficient proof — if one was needed — that the Corps never showed any discrimination in the selection of officers and men for service in India. Major (Brevet Lieutenant Colonel) M. Ṣ. Brander and Captains C. M. Smith and H. C. Goodfellow became Directors of Supplies and Transport; Major E. H. Fitzherbert became Inspector of the Royal Army Service Corps and Captain E. B. Rowcroft became the first Director of the Royal Electrical and Mechanical Engineers. Captain A. T. F. Smith was later DEME of the Indian Army. Captain R. I. Jones, Indian Army Service Corps, who was Director of Supplies and Transport Indian Army from 1946 to 1947, was also serving at Chaklala at the same time.

By the end of May 1928 all the Royal Army Service Corps officers and men who were not transferring to the Indian Army Service Corps had left India, except those who held appointments on the Staff, some returning home, others, who still had part of their overseas tour to complete, like Major E. H. Fitzherbert, Captain C. M. Smith and Captain H. C. Goodfellow, moving on to Hong Kong and Shanghai and other foreign stations. Among those who had served in India there was genuine sorrow that ten years of close and friendly collaboration between the two corps had come to an end.

The final act was the publication of the Army Order in the autumn of 1928 announcing that His Majesty the King had approved the disbandment of the twenty two Royal Army Service Corps Indian MT companies, on their replacement by similar units of the Indian Army Service Corps.

Many tributes were paid to the achievements of the Royal Army Service Corps in India. When he unveiled the memorial erected at Chaklala to the 'Memory of the officers, warrant officers, NCOs and men of the Royal Army Service Corps who have died in India' on 5 January 1927, General Sir Alexander Cobbe VC KCB KCSI DSO, General Officer Commanding in Chief Northern Command, referred to the impending transfer to the Indian Army Service Corps of the management of Mechanical Transport in India, and explained how necessary it had been for the Royal Army Service Corps to take on this commitment and how efficiently it had been carried out.

Major General ap Rhys Pryce (later General Sir Henry ap Rhys Pryce KCB CMG DSO, and Master General of the Ordnance, India), who succeeded Major General Wilson as Director of Supplies and Transport in India in 1925, said in his farewell speech at the RASC officers' mess, Chaklala, on 24 October 1927:

'I will take this opportunity when the RASC in India is, in the course of this winter, going to hand over the mechanical transport to the IASC, to say a few words, firstly, of welcome to those of the RASC, who are joining the IASC, and secondly, of farewell and thanks to those who are going home and not joining the IASC. In bidding farewell to those of the RASC who are leaving India, I would thank those now present and those already gone for all the good work that they have done in India.' And he pledged himself and the Indian Army Service Corps to cherish and care for the memorial which had been erected at Chaklala.

In an Indian Army Order (No 337) of May 1928, Sir William Birdwood, the Commander in Chief, bade farewell to the Corps with these words:

'On its departure from India, His Excellency the Commander in Chief in India wishes to place on record the deep obligation of the Army in India to the Royal Army Service Corps.

'During the Great War, it was necessary to increase the scope of the mechanical transport then in India, and the Imperial Government were requested to send out representatives of the Royal Army Service Corps to carry out the work. The members of the Royal Army Service Corps who came to India have worked with such loyalty that they have brought into being a service which the Indian Army Service Corps will be proud to maintain.

'In doing this, they have carried out the traditions of their Corps at the cost of nine officers and one hundred and sixty-three other ranks who have died in this country.

'In saying farewell to the Royal Army Service Corps, His Excellency takes the opportunity to welcome those members of the Royal Army Service Corps transferring to the Indian Army Service Corps.

'They are coming into a Corps which is imbued with the same traditions as that which they have just left. Those traditions are "Service for others, whatever the risks that may be entailed" '.

A year later, in a letter to the *Journal* of The Royal Army Service Corps, Field Marshal Sir William Birdwood wrote:

'As the senior officer on the active list of the Indian Army, I fully realise the advantages and indeed the honour of the Indian Army Service Corps being now so very closely associated in every way – both in its work and in sentiment with the RASC. I can say with confidence that the same spirit that has always actuated the RASC is also fully alive in the IASC, and for all practical purposes, as regards service in the field (for which the Army exists), the two services may be regarded as one.

'I write these few words from one of our out posts at the extreme north-west corner of the Indian frontier, and where during the last few weeks I have had every opportunity of seeing some of the wonderful work done by the IASC, helped, as they have been, to a very great extent, by officers and men of the RASC.

'It gives me the greatest pleasure to be able to write these very few words regarding the service for which I have the very greatest respect and admiration.'

On 3 June 1935, the title Royal was conferred on the Indian Army Service Corps by His Majesty The King in the Silver Jubilee Birthday Honours. The following exchange of telegrams took place:

'DST, SIMLA, – RASC heartily congratulate RIASC on grant of Royal title. – MACLEOD.'

'GENERAL MACLEOD, WAR OFFICE, – All ranks RIASC thank the RASC for kind message of congratulation. – STEWARD, SIMLA.'

A Rolls Royce Armoured car as used for patrols in India in 1922.

ABOVE: A typical stretch of hill road in the arid country between Kohat and Peshawar, the North West Frontier Province, India. The leading vehicle is a model-T Ford. BELOW: RASC units parked in 1923 at Razani along the circular road linking the Garrisons of Wazir and the Mahsud Territories.

In the background is the country of Afghanistan at the end of the Khyber Pass; in the right middleground is a camel convoy wending its way along the old road; in the middle/foreground is a convoy of Karrier RASC vehicles travelling in packets of three, and in the left foreground can just be seen the derelict standards of the Khyber Ropeway — c1925.

ABOVE: One of the forts in the Tochi Pass at Miranshah, c1926. The vehicles being tested are (left) one of the original medium 6-wheeled Thornycroft and (centre) one of the early 6-wheeled Morris. BELOW: The original War Office type medium 3-ton 6-wheeler, built to War Office Specification by Thornycroft, c1927 in Rawalpindi, Punjab India. This prototype proved inadequate for Indian conditions: the twin rear wheels were no good in deep sand, and the cooling was inadequate. Eventually, a modified Indian specification was adopted. Major C. B. Evans RASC and Major R. K. Hubbard RASC were concerned with the new specification.

LEFT: Major General E. H. Fitzherbert CBE DSO MC, in 1926, as Lt Colonel commanded the Razani Force; he later became Inspector RASC. RIGHT: Major General E. W. Rowcroft KBE CB was to become the first Director of the Royal Electrical & Mechanical Engineers and was one of six RASC officers serving in Chaklala in 1927 who later rose to the rank of Major General. FAR RIGHT: The GSM with three bars — Afghanistan NWF 1919; Waziristan 1919-21; and Waziristan 1921-24 — gained by Lt Col J. C. C. Cooke. BELOW: A 3-ton 6-wheeler Karrier, c1927; this vehicle was much used in India and sold to civilian contractors to clear the ruins of Quetta, after the severe earthquake.

Local resources being enjoyed by men of the Rhine MT Company RASC, in the Cologne brewery in which they were billetted.

THE BRITISH ARMY OF THE RHINE *VI*

When the British Army of Occupation first arrived in Germany at the end of 1918, it was a large force scattered over a wide area. 2 Corps Troops MT Company Royal Army Service Corps played a leading role in supporting the move into the Rhineland. The number of units was gradually reduced and concentrated into an area with its centre at Cologne. The force included an Air Force squadron and a detachment of the Royal Navy in the shape of the Rhine Flotilla.

In the early days there were constant adjustments to unit locations due to disbandments and demobilisation. By the autumn of 1920 the Royal Army Service Corps was represented by a small directorate at headquarters of the British Forces on the Rhine, consisting of an Assistant Director of Supplies and Transport, in the rank of colonel, and two deputies, a Deputy Assistant Director of Supplies and a Deputy Assistant Director of Transport, and seven units. These were: Rhine Forces (MT) Depôt, commanded by a lieutenant colonel, Rhine Forces Train, Rhine Forces (MT) Company, 18th Army Auxiliary (H) Company, commanded by majors, and 498 (MT) Company, commanded by a captain.

The Rhine Army consisted of one Cavalry regiment, eight Infantry battalions, two Royal Artillery units, a company of the Royal Tank Corps, HM Rhine Flotilla, one Royal Air Force squadron, Royal Engineer units and ancilliary services.

Bulk supplies were sent from France and England to Antwerp and Rotterdam, and from there by rail or barge to Cologne. In the early days of the occupation there were not many local resources available. The Military Government officials were concerned that civilian needs should have first call on available stocks.

Rhine Forces Train was originally the 41st Divisional Train, re-named London Divisional Train, soon after Rhine Army took over its duties. Not long after, it was reduced by one brigade company, leaving only three companies.

At first the Royal Army Service Corps was kept busy. There was a great deal to be done in restoring conditions to normal and transport was at a premium. In addition to garrison duties there were many demands from Military Government, whose officials were no more popular here than they have been elsewhere, either to their fellow servicemen, or to the local population, and there were still Foreign Office commissions all over Europe. During 1921 and 1922 there were strikes among German workers which added to the load.

Plebiscites were being held in a number of frontier areas and all these required military forces to supervise the voting and to prevent intimidation. In all these the RASC played its part.

On 26 January 1920 a small advance party arrived in Danzig from Cologne to start taking over the Free State. By the 30th a supply section and an MT detachment RASC had arrived. On 8 February Sir Reginald Tower, the Allied Administrator, took over the Free State from the Germans. Lieutenant General Sir R. C. B. Haking KCB KCMG was in command of the

Allied troops. At GHQ there was an Assistant Director of Supplies and Transport (Lieutenant Colonel E. T. Hills OBE) with a staff of eleven. Lieutenant Colonel Bingham RASC was appointed AA and QMG Plebiscite Area and Danzig in April.

By November the plebiscite was completed and, after the Order in Council creating Danzig a Free City had been issued, evacuation was ordered; by the first week of December the force was back in Cologne. At first the population had been sullen and uncooperative but the inherent friendliness and lack of malice of the British soldier, as usual, made its impact and, when the time for evacuation arrived, the Danzigers were almost sorry to see them go.

There was a small RASC detachment in Berlin as part of one of the missions, headed by a lieutenant colonel. There were two missions in Berlin, one mainly engaged in the repatriation of prisoners-of-war, and the other, the Inter–Allied Mission of Control, which was appointed to deal with certain terms of the peace treaty. The complete commission did not arrive until after the treaty was ratified. Small detachments were then sent all over Germany. One of the achievements of the British Force in Berlin was to persuade the Berliners to take up 'Der Kricket' and one of their matches was actually reported in *The Times*.

The other plebiscite in which the Royal Army Service Corps was involved was the Inter-Allied Plebiscite Commission in Upper Silesia, on which Colonel Sir Harold F. P. Percival KCMG CBE DSO was the British representative from August 1919 until the Commission's work was completed. A large detachment of 585 (MT) Company, from Cologne, was in Upper Silesia during the plebiscite. Colonel Percival had been adjutant at Woolwich, a staff captain in the Transport Directorate of the War Office at an important time of reorganisation before the war, and then an instructor at the Staff College. As a result of the plebiscite, Upper Silesia was partitioned between the new state of Czechoslovakia, Poland and Germany, most of the richest mining areas going to the two former states. On the completion of the commission's task in September 1921, Colonel Percival returned to the Rhine Army. He later became Assistant Director of Supplies and Transport in Malta, until his retirement in 1929.

Just over seven months after the Armistice, peace was signed at Versailles on 29 June 1919, the treaty ratified on 10 January 1920. The war was at an end, but this was not the end of troubles in Germany. Until the peace treaty was signed and ratified the Allies were unwilling to reduce their occupying forces. The French were determined to see that the terms of the Treaty were rigorously enforced and insisted that reparations should be met in full. In the prevailing state of German industry this was impossible without crippling the entire economy. The situation was not improved by the unco-operative attitude of Germans towards what her statesmen regarded as a harsh and vindictive treaty, which they described as the 'Diktat' of Versailles. At the end of 1922, when the Germans were in default over payment of reparations, the French Government unilaterally ordered their troops to occupy the Ruhr, and for eighteen months a rigorous regime was imposed on Germany's richest industrial area.

The inevitable result of the French attitude was the complete collapse of the German economy. The state of the mark at the end of the war was precarious. Its fall in 1923 was catastrophic. It fell from a thousand to the pound to a million and then to a billion. At a stroke savings and pensions were wiped out. Salaries lost their value overnight. Inflation on this scale made money meaningless. Property and possessions were the only things which retained any value. 1923 witnessed the heyday of the black market and a return to the system of barter.

But the situation could not continue. By accepting enormous sacrifices Herr Stresemann, the new German Chancellor, persuaded the Allies to help Germany stabilise the mark. At a rate of one Rentenmark to one billion marks a new currency was introduced by Herr Schacht, the Finance Minister. In the new atmosphere of co-operation the Dawes Plan was evolved and approved in March 1924, which, in exchange for a revised reparations settlement, provided foreign loans for the reconstruction of Germany's industries.

The years 1924 to 1929 saw a steady improvement in relations between the occupying forces and Germany, and a gradual reduction in the Rhine Army forces.

Meanwhile, the titles of Royal Army Service Corps units had been rationalised to fit the establishments of home and overseas stations and, by the end of 1924, the Royal Army Service Corps units consisted of 4 and 5 (HT) Companies, 28 and 29 (MT) Companies and the Rhine Supply Depôt.

On 10 July 1925, when the Secretary of State for War paid a visit to the British Army of the Rhine, 5 (HT) Company took part in the review parade at full strength with 12 pairs of horses and 12 GS wagons. In the autumn of that same year, brigade training took place for the first time in Germany.

To make good the shortage of mechanical transport, German-hired lorries were used and a mobile force, based on 3-ton Albions, survivors from the war, was a feature of the exercise. 'Tactical transport' was used in a troop-carrying role and horses were carried three to a lorry.

At the end of 1925 it was decided that the Rhine Army would move from the Cologne area to Wiesbaden and 29 MT Company was the first RASC unit to go, on 1 December. By early in the new year the move had been completed. At the same time a new unit, 34 (MT) Company (Mobile Repair Unit) RASC, was formed at Wiesbaden, and the Rhine Supply Depôt was re-named 'E' Supply Depôt, to bring it in line with other RASC supply depôts. 4 (HT) Company, 28, 29 and 34 (MT) Companies and 'E' Supply Company were all concentrated in the newly named Buller Barracks at Wiesbaden. 5 (HT) Company shared part of the Royal Artillery barracks at Biebrich, some three miles from Wiesbaden. The commanding officer now was Lieutenant Colonel C. W. Horton DSO, the Assistant Director of Supplies and Transport, Colonel R. P. Crawley DSO OBE MVO and the District Barracks Officer, Colonel J. C. Browne CMG DSO.

Later, in 1926, 4 (HT) Company was withdrawn from the Rhine Army to Catterick, which enabled all the Royal Army Service Corps units to be accommodated together in Wiesbaden. Two years later, in February 1928, 5 (HT) Company was also moved from Rhine Army to Aldershot, leaving no Royal Army Service Corps horse transport units in Germany.

Under Herr Stresemann, Germany's industrial recovery continued. By 1926 her output was only 5% below that of pre-war years. In September of the same year Germany was admitted to the League of Nations by the Locarno Pact, but under clause 131 of the Versailles Treaty she still carried the responsibility for the Great War. Nevertheless, the Inter-Allied Control Commission was withdrawn from Germany in January 1927. When the representatives of fifteen nations, including Germany, signed the Kellogg Pact for the outlawry of war on 27 August 1928 and later ratified it in 1929, it really seemed as if Germany had 'come in from the cold'.

The first decade had passed without hostilities, since the declaration of the policy based on the assumption that there would be no major war in the next ten years. It had been necessary to undertake a number of imperial policing operations, but these had all been on a small scale, and none of them had had serious consequences; on the contrary, they had helped to alleviate the tedium of peace–time soldiering with inadequate resources, and to point out to the politicians that there was still a need for sufficient forces to carry out our commitments. It could well be claimed that that policy was vindicated.

On the tenth anniversary of the signing of the Treaty of Versailles it appeared to most people that the world was moving towards a period of peace and stability. The material damage done by the war had mostly been repaired. Germany had recovered from the disastrous financial collapse of 1923 and her economy, unfettered by defence expenditure, was rapidly expanding; she seemed to be ready to take her place among the community of nations. The United States had enjoyed a post-war boom of unparalleled prosperity and standards of living were rising everywhere. A soldier stationed on the Rhine, describing the German national memorial, situated on the Niederwald Heights overlooking the Rhine at Bingen-Rudesheim, wrote:

'With the growing prestige of the League of Nations, Locarno and the peace pact, and the continued efforts by the world's statesmen to find a peaceful solution of international disputes, it may be that the Rhineland has at last seen the ghost of war laid low, and that "Germania" may henceforth be but a reminder of days when man resorted to primitive means of redress.'

And in December 1929 the Editor of the *Journal* of the Royal Army Service Corps could write: 'It must be comparatively rare in the annals of the British Army that the period of Christmas finds us really at peace. Those of us who escaped the war-like expedition to Shanghai a year or so ago had visions this season of eating Christmas pudding in genteel fashion from the heights of the Wailing Wall. Fortunately, we have been spared. No sign of war-clouds seems to be apparent on the horizon. Our age-long foe, Financial Stringency, is with us to remind us how much at peace we really are.

'By the time this number appears in print, the Rhine Army will be no more . . . The disappearance of our Army on the Rhine marks a definite stage of progress in European goodwill.'

Even Churchill recorded that, when he left office in 1929 he felt hopeful that the peace of the world would be maintained. Few could foresee that the world was on the edge of one of the greatest economic depressions ever experienced, or that the growth of Fascism in Italy or the conflict in Germany between Communism and National Socialism, with its doctrine of Aryan supremacy and hatred of the Jews would, in the space of a few years, lead to an unbridled resurgence of military ambition and megalomania.

And so, in 1929, the decision was taken to withdraw the British Army of the Rhine. 29 (MT) Company was destined to be established at Aldershot as a lower establishment company, equipped with workshops manned largely from 34 (MT) Company (Mobile Repair Unit). 28 (MT) Company, as a higher establishment company, was to go to Catterick, accompanied by 34 Company, which was to be reduced to the home establishment of a workshop company. 'E' (Supply) Company was to be disbanded, after having performed the role of 'parent' of the 150 or so details of the Royal Army Service Corps, who would form part of the rear party.

The departure of the first batch of Corps families from the Rhineland was witnessed by a large crowd at Wiesbaden main station on 23 September, as the special train provided for the 2nd Battalion The Leicestershire Regiment, with special coaches for Corps families, started on its homeward journey. 28 (MT) Company and 34 (MT) Company had arrived at Catterick by the end of November.

Throughout the occupation, Rhine Army was treated as a home command. It was a popular station, with plenty of opportunity for sport and recreation. Barracks and married quarters were comfortable, and the facilities in Cologne and later, Wiesbaden, were better than in most other stations.

'The Corps on the Rhine are as happy a family as at any Corps station in the world', wrote the adjutant (a future Inspector of the Royal Army Service Corps). And he concluded with the words: 'These men of the little Rhine Army have been good ambassadors in the interests of European goodwill. On the last coach of that special train there was chalked, in German handwriting "Auf Wiedersehen" — "See you again." '

How little did anyone realise that only sixteen years later that wish would be fulfilled!

The final act in the evacuation of the Rhineland took place on 30 June 1930, when the French, somewhat reluctantly, withdrew their forces of occupation. At the same time the Inter-Allied Rhineland High Commission ceased its activities. The Germans were still bound by the Treaty of Versailles and the Locarno Pact to treat the Rhineland as a demilitarised zone, but they were quietly jubilant at the removal of the occupying forces from the 'Vaterland', and celebrated the occasion by over-printing the current postage stamps, depicting the head of Hindenberg, with the date '30 Juni'.

ABOVE: RASC Company marching to church, Wiesbaden, 1928. (MHY) BELOW: Buller Barracks, Wiesbaden, Germany. (MHY)

Children of Shanghai, 1927. (BC)

Shanghai *VII*

Shanghai was first settled by Europeans under the terms of the Treaty of Nanking in 1842, after the Opium War. Hong Kong was ceded to Britain at the same time.

Since the first settlements in 1843 the areas under foreign control were subject to periodical threats of attack, either from regular Chinese troops, tempted by the riches of such a prosperous area, or from rebels against the existing Chinese regime. No Chinese authority was prepared to give protection to the foreign settlements and it was soon realised that they must look after their own defence. Over the years they had established three special rights, tactitly recognised by the Chinese authorities: (i) The right to organise their own defence force and to call on the assistance of the naval and military forces of foreign powers; (ii) The right to preserve a state of armed neutrality in civil wars, and (iii) The right to exclude Chinese armed forces.

Ever since the Republic of China was established in 1912 the country had been in a turmoil. Strong opposing factions developed in North and South. An independent Republic of South China was declared by the southern armies, while in the North parliamentary government was abolished. Meanwhile, a Nationalist Party with a traditional anti-foreign policy was springing up, soon to become well known as the Kuomintang. Anti-Japanese riots broke out in Shanghai, in the summer of 1925, followed by a general strike, and fighting took place between rioters and the crews of Japanese and European warships. Before the end of the year civil war had broken out in China. A year later the Southern Nationalists were advancing from Hankow towards Shanghai and there were fears that the city might fall into their hands, with the consequent risk of danger to British lives and property in the International Settlement Hong Kong was boycotted for more than fifteen months.

Tientsin also felt the effects of the civil war. Minor hostilities between opposing war-lords for personal gain, or to satisfy private spite and revenge, were almost continuous in Northern China. Towering above the lesser belligerents were Marshal Chang-So-Lin, an ex-bandit and Feng-Hu-Hsiang, the 'Christian General'. But, although troops were frequently put on the alert, these activities had little effect on the foreign communities in Tientsin except for occasional delays in the delivery of mail.

When anti-European riots broke out again in January 1927, followed by another general strike in Shanghai, and the conflict between the Cantonese and the northern troops had reached a point no further away than the Tsientang River, the foreign powers decided that the time had come for stronger measures.

Shanghai, straddling the Whangpoo River, a tributary of the mighty Yangtze-Kiang, is the commercial gateway to China. It has a larger geographical area to serve and a more important hinterland than any other port in Asia, and is one of the eight largest ports in the world. Ample berthing facilities exist for the biggest ships and enormous quantities of produce from the interior arrived in Shanghai for shipment overseas. Foreign investment made the city one of the richest in the world, with all the amenities and resources which justified its title as 'The Paris of the East', contrasting everywhere with characteristic Oriental squalor.

In the mid-twenties the urban population was already more than a million, which included some 40,000 foreigners, of whom about 8,000 were British, and a large proportion Japanese. Shanghai consisted of four main areas, roughly divided by ill defined lines running east to west.

In the north was Chapei, an entirely Chinese area. Next came the International Settlement, then the French Concession, and in the south was the Chinese City. Chapei and the Chinese City were governed by the Chinese Municipality of Greater Shanghai. The French Concession was entirely under French control and its defence was largely in the hands of French colonial troops. The International Settlement was governed by a Municipal Council, which was subject to the final authority of the Court of Consuls. The Municipal Council was elected annually, and a typical composition was five British, one American, one Danish, two Japanese and five Chinese members. The staffs of the various municipal services were largely British. The International Settlement maintained an International Volunteer Force, about 1,700 strong, with a British colonel as commandant and a regular British officer as adjutant. Included in the Volunteer Corps was a force of about 300 Russians. The international settlements were built on what were once mud flats and were of varying sizes. The British Concession was about eight square miles in extent and the French Concession, which bordered it, about four square miles. In the centre of the city was the race course, a wide expanse of some seventy acres, which provided opportunities for all sorts of recreation.

Such was the place which was to become familiar to all those who found themselves part of the Shanghai Defence Force, or 'Shaforce', which was despatched to the Far East at the beginning of 1927.

In early January of that year two infantry battalions and a proportion of the Mediterranean Fleet were despatched with all speed from Malta. On Saturday 22 January orders were received at Aldershot for the mobilisation of a supply depôt, a bakery detachment and other details for overseas service. By the 28th the draft was ready and, after inspection by Lieutenant Colonel W. N. White DSO, Commanding Officer of Service Companies, it entrained at the government siding under the command of Captain H. B. B. Butler OBE, who had Lieutenant E. D. Dynes and Lieutenant D. H. V. Buckle with him, and sailed for China in the *Kildonan Castle*.

On the same day, 12 (MT) Company at Woolwich received its orders for duty with the Shanghai Defence Force. The company, under the command of Major G. E. Toynbee CMG CBE, embarked at Southampton in HT *Assaye* and at Liverpool in SS *Bellerophon* (which took the vehicles) between 26 January and 6 February. The strength of the company was seven officers and 174 other ranks, with 49 vehicles, six motor-cycles and two trailers. It was reported from Woolwich that other units regarded the fate of 12 Company with envy.

As part of the force was to include an Infantry Brigade from India, orders were also received by 3 (Indian) MT Company RASC (Motor Ambulance Convoy) at Chaklala to proceed to China, and on 27 January the unit left Rawalpindi station for Bombay, under the command of Captain A. F. St C. Collins OBE MC, who had Captain H. J. Walsh as his workshop officer. By 22 February they arrived at Shanghai with their 20 Fiat ambulances, as the first mechanical transport company to serve in China.

Another S and T unit from India was the 18th Indian Divisional Troops Transport Company, which consisted of more than 400 mules.

Finally 38 (MT) Company, specially formed for the purpose, and equipped with the latest six-wheeled Morris 30-cwt lorries,and a supply unit were despatched from Aldershot on 8 April 1927. Included were some 100 Reservists, specially called up for the emergency. Before entraining the contingent was inspected by the General Officer Commanding Aldershot

Command, Lieutenant General Sir David G. M. Campbell KCB, who was accompanied by Colonel Commandant E. Gibb CB CMG CBE DSO, Assistant Director of Supplies and Transport. The GOC stressed that the trouble in China was entirely domestic and sending troops there was not to be construed as a desire to interfere in wars waged between rival Chinese factions. Our troops were going solely for the purpose of protecting the lives, property and interests of British subjects.

The contingent amounted to 226 all ranks, under the administration of Bt Lieutenant Colonel G. E. Badcock CBE DSO as OC RASC. Captain S. W. Morrison was in command of 38 Company and Lieutenant C. A. Hunt, of the Supply Unit. These units, which formed part of the 15th Infantry Brigade, sailed in SS *Derbyshire* and SS *City of Marseilles* from Southampton on 19 April 1927.

The international force included British, French, American, Dutch, Italian and Japanese naval units, a French army contingent, an Italian force and the 4th United States Marines, who were conspicuous in the streets of Shanghai. The British Army force, apart from the RAF detachment, numbered around 15,000 at its highest figure and consisted of the 9th (Jansi) Brigade of the Indian Army and the 13th, 14th and 15th Infantry Brigades. In addition there was the 12th Royal Marines. 'Shaforce' was strongly backed up by the Shanghai Volunteer Corps, a first class territorial force, and the Shanghai municipal police.

Initially there were accommodation difficulties in Shanghai and some units had to be temporarily held in Hong Kong, in camps in Kowloon under canvas. In fact, 38 (MT) Company never reached Shanghai as a formed unit, and most of 15th Infantry Brigade was still in Hong Kong in September, when a start was made to reduce the force.

The Force was commanded by Major General J. Duncan CB CMG CVO DSO, who later commanded the 1st Division at Aldershot, and he was succeeded by Major General A. E. Wardrop CB CMG. Command Headquarters was in the Kalee Hotel. The senior Royal Army Service Corps Officer was Lieutenant Colonel D. C. Cameron OBE, Assistant Director of Supplies and Transport who, on 23 July was promoted Colonel and appointed Assistant Director of Supplies and Transport, North and South China Commands.

On arrival, 12 (MT) Company was accommodated in a tramway repair depôt at Baikal Road, about a mile from the town which, with the usual ingenuity of a Royal Army Service Corps unit, soon became one of the best quarters in Shanghai. The supply depôt and bakery were at Old Ningpo Wharf on the Whang-Poo River, where the Indian supply unit, which had arrived first, had established itself.

During March and April numerous reinforcements arrived and life settled down. Full use was made of the race course and other open spaces for recreation. An impressive King's Birthday Parade was held with the colour of the 2nd Battalion, The Coldstream Guards being trooped, and an ambitious torchlight tattoo was staged. The local population was generous in entertaining and providing facilities for the troops, and excellent relations were established with the Shanghai municipal police and the Shanghai Volunteer Corps.

A convalescent depôt and change-of-air camp was established at Wei–Hai–Wei and a small Royal Army Service Corps detachment was required to help with administration. Normally the camp only operated during the summer months, but a small rear party remained on a care taking basis during the winter. Wei-Hai-Wei is the port of a small tract of China, a spur of land in the north-east corner of the province of Shantung with an area of 285 square miles and a population of 170,000, leased to Great Britain under the Convention signed at Peking on 1 July 1898 and ratified in London on 5 October 1898. It is about 450 miles north of Shanghai and made a delightful rest station for the summer and autumn months, with perfect weather, beautiful scenery, bracing air and excellent bathing. Wei-Hai-Wei was opened again in 1928 as a summer camp and again in 1929, but then only as a convalescent depôt. This, the most easterly station maintained by the Royal Army Service Corps, was handed back to the Chinese Government on 2 October 1930 after 32 years of British sovereignty, in fulfiment of a promise made at the Washington Conference of 1921. So ended a period of peace and quiet in this haven beyond the border of Shantung, one of the most lawless and bandit ridden provinces of China.

Meanwhile, in Kowloon, in company with the rest of 15th Infantry Brigade, known as 'Reforce', 38 (MT) Company with their 60 Morris six-wheelers and the mules of the 18th Divisional Troops Transport Company IASC, occupied themselves in training and putting the Morris 30-cwt lorries to all manner of unusual tests, most of which they surmounted triumphantly. Weekly demonstrations of their capabilities were arranged near Fanling in the New Territories, which aroused tremendous interest among members of other units and the civilian population. Lieutenant Colonel Badcock combined the duties of OC RASC South China Command and OC RASC 15th Infantry Brigade.

The strong force which had been sent to Shanghai had had its effect, and by August 1927 it was clear that it could be reduced. The troops from India, having been the first to arrive, were the first to go, with the Durhams and Glosters of the 9th (Jhansi) Infantry Brigade in the lead, followed by the rest of the brigade and the other units from India, including 3 (Indian) MT Company RASC (MAC) and a detachment of 18th Divisional Troops Transport Company, IASC.

3 (Indian) MT Company embarked at Shanghai in the *Bellerophon* on 11 September 1927 and arrived back at Sitapur, United Provinces, on 30 September, all the better for six months experience of service in China during the critical days in Shanghai.

With the departure of these units, part of the 15th Infantry Brigade started to move up from Kowloon during September and October. In November it was decided that some of the officers and men of 38 (MT) Company would be required as relief for personnel of 12 (MT) Company in Shanghai; all the Corps Reservists of 38 (MT) Company and the Supply unit of 15th Infantry Brigade, and all the vehicles of 38 Company, would return to Woolwich, sailing in HT *Bellerophon*, while the remainder would sail in the next troopship, the *Somersetshire*, in mid-November. Among them were Lieutenant Colonel Badcock, Captain Pessell and Lieutenants Dynes and Buckle. Also in the *Bellerophon* were the 420 mules of the 18th Indian Divisional Troops Transport Company on their way back to India. The 12th Royal Marines, who had been landed at the beginning of the emergency, left Shanghai on 6 December 1927.

By 14 February 1928, with the sailing of HT *Assaye*, the force had been reduced to four battalions, and by the end of the trooping season, just over twelve months after the troops' arrival, the situation was beginning to stabilise. 14th Infantry Brigade Headquarters became Shanghai Area and was located in the same building as Command Headquarters.

In April there were two new arrivals from India, as a result of the transfer of responsibility from the Royal Army Service Corps to the Indian Army. Major E. H. Fitzherbert DSO MC arrived on the 17th for duty in the ST Office, and later in the year became acting Assistant Director of Supplies and Transport; on the 28th Captain C. M. Smith MC arrived to join 12 (MT) Company.

In July Colonel D. C. Cameron was appointed Commandant of the Royal Army Service Corps Training College, after a year and a half of strenuous activity as the senior Royal Army Service Corps officer in a command which extended from Hong Kong in the south to Tientsin and Peking in the north. His successor was Colonel H. A. Stewart DSO OBE from Aldershot (a direct exchange of appointments) who, however, did not arrive until several months after his predecessor had left.

Apart from creating a presence in Shanghai itself, the force was involved in minor expeditions to protect European industrial enterprises in China, such as 'Treforce', organised for the protection of the British-owned mines of Tongshan in the late summer of 1928. The Corps always had a part to play in these excursions.

By the time that Colonel Stewart arrived on 6 November 1928, the strength of the Royal Army Service Corps in Shanghai was only 228. Major Fitzherbert was appointed OC RASC on 1 December — the first officer to hold this appointment in Shanghai. Other changes were also made. There would continue to be a garrison in the city of Shanghai; China would be a three-year station, of which only twelve months would be spent in Shanghai, which would not be a family station. At midnight on 12/13 April 1929 North and South China Commands were united, with the headquarters of the new China Command in Hong Kong. Henceforth, Colonel Stewart, the Assistant Director of Supplies and Transport, would be in Hong Kong. Headquarters Shanghai Area remained the only headquarters in Shanghai.

Life became much easier and full use could be made of the recreational facilities of the race course. Time was found to form a branch of the Royal Army Service Corps regimental association and 12 (MT) Company could boast 100% membership. Captain C. M. Smith, who had been acting Deputy Assistant Director of Transport in the absence of Captain Collins, was warned for service in Mauritius and the local correspondent of the *Journal* of the Royal Army Service Corps commented:

'Captain C. M. Smith MC, who now commands the Company (12 MT Coy), will be leaving us shortly for the sunny clime of Mauritius, where the strength of the RASC appears to be one officer, one warrant officer, two staff sergeants, one lance corporal, two mules and an ox. We noticed him in earnest consultation with Sergeant Blenkinship over Infantry Training, Vol I in an effort to improvise a detail for forming up his new command on parade.'

A little more than two years after the emergency had begun the situation in Shanghai was, for the time being, settled. The United States Marines had left and naval forces were greatly reduced. It was still necessary to keep a small garrison there but, with the reduction of one more battalion, leaving only two, the strength was only a little more than 2,000. Of this only 183 belonged to the Royal Army Service Corps — 130 in 12 (MT) Company, 25 in the supply depôt and bakery, and 28 headquarter clerks and other details.

The whole garrison was moved to the Western District of Shanghai. Headquarters Shanghai Area left Kalee Hotel, which was soon demolished to make way for new buildings, for Yu Yuen camp. The supply depôt and bakery were moved from Old Ningpo Wharf to Ash Camp, and 12 (MT) Company from the Bikal Road tramway depot to Tifeng Road Camp.

Let Major Fitzherbert have the last word:

'One final picture. The Royal Army Service Corps are playing the Shanghai police. It is nearly 6 o'clock. We have been "outed" for under a hundred and but for an heroic innings by Lieutenant Shaw, should have made much less. The police have already passed our score with eight wickets in hand, and are thoroughly enjoying themselves. The writer is the bowler, a significant indication of our plight. He pauses at the end of his seven yard run (McLaren only takes seven yards). What has happened to the fielders? They appear to be taking a deuce of a time to change over. Another glance round. Straight down the wicket comes a Chinese amah wheeling a perambulator. From between first and second slip appears a man with a wheelbarrow containing ten Chinese girls. The obsolete position of long stop is occupied by a Chinese lady with an infant taking nourishment at Nature's fount. Mid on appears to be a coolie with a bamboo pole, from each end of which dangles a basketful of live ducks. Behind the bowler's arm a rickshaw coolie is plying for hire, while the square leg umpire is having his attention distracted by a most persistent beggar, asking for alms. All the world seems hurrying westwards. Apparently the Scots Guards drums are about to beat "Retreat". It is time to draw stumps.'

But although the Emergency was over and, for the next four years, life in Shanghai was comparatively uneventful, the garrison was retained. The newsletter published in the *RASC*

Journal recorded such peaceful matters as the opening of a new officers' mess in the autumn of 1930 and the closing of the site at Hayims Garden, when the move took place with no more excitement than the wrecking of Captain Mead's ukulele and the safe journey of Captain Craig-McFeely's menagerie of five canaries and a lady dog.

Early in 1932 troubles broke out again in Shanghai. These were the result of disputes between the Japanese and Chinese, culminating in an anti-Japanese boycott in Shanghai and student riots.

On 28 January 1932 a state of emergency was proclaimed by the Municipal Council and a curfew imposed from 10.30 pm to 4 am. The defences of the International Settlement were manned and the Shanghai Volunteer Corps mobilised. Moving to their defensive positions north of the International Settlement, the Japanese forces encountered resistance from the troops of the Chinese 19th Route Army. In order to protect their nationals, the Japanese landed more than 50,000 troops and there followed five weeks of hostilities, with heavy shelling and bombing by the Japanese, and the virtual destruction of a large part of Chapei. Although the other powers brought in reinforcements, they were not involved and peace was soon restored.

At the end of 1933 orders were given for the disbandment of 12 Company and the dispersal of the company's personnel to various stations, mainly Hong Kong. In February 1934, 12 (MT) Company, which had been in Shanghai since 24 March 1927, was nominally transferred to Hong Kong, where it was re-formed under the command of Captain L. J. Walch. This left the strength of the Corps in Shanghai in the summer of 1934 as five officers and forty-five Warrant officers, non-commissioned officers and men. The station now became a three-year tour of duty for officers and in June 1935 the first married families arrived.

It was not long before another emergency occurred. In August 1937 the Japanese decided to take reprisals for the shooting of two Japanese nationals by Chinese soldiers. Japanese warships appeared off Shanghai. The Shanghai Volunteer Force was mobilised and the International Defence Force manned the defences. On 14 August bombs were dropped by the Chinese inside the settlement. On 16 and 17 August two infantry battalions arrived from Hong Kong and two days later British women and children were evacuated to Hong Kong. On 23 August there were 600 casualties in the settlement. Under pressure from the Japanese, the Chinese withdrew westwards, and by November hostilities had moved away from Shanghai. On 26 November families returned from Hong Kong.

ABOVE: Shanghai waterfront, 1925. LEFT: Thibet Road, Shanghai, a street in the French Concession and RIGHT: Mohawk Road, Shanghai, a street in the British Concession — both in 1927.

ABOVE: Shanghai Volunteer Corps armoured car patrols streets with 2 Bn Durham Light Infantry. BELOW: Shanghai bird fanciers in 1927. (BC)

ABOVE: A barrack room of 12 Company RASC in Shanghai in 1930.
BELOW: 12 Company marching past the General Officer Commanding Shanghai in 1930.

DETAILS OF THE DE

The total number of officers and men of all nations (naval and military) at p neighbourhood of 40,000 (excluding Municipal Defence Forces), of which Great Britain co

BRITISH

Royal Navy

COMMANDER-IN-CHIEF

Rear-Admiral Sir Reginald Tyrwhitt, Bt., K.C.B., D.S.O., D.C.L.

Army

GENERAL OFFICER COMMANDING

Major-General John Duncan, C.B., C.M.G., C.V.O., D.S.O.

GENERAL STAFF

G.S.O. 1—Colonel Viscount Gort, V.C., D.S.O., M.V.O., M.C. (Grenadier Guards).
G.S.O. 2—Major the Hon. P. G. Scarlett, M.C. (The Buffs).
G.S.O. 2.—Lt.-Col. T. G. G. Heywood, O.B.E. (R.A.)
G.S.O. 2.—Major C. D. Rawson (R.A.)
G.S.O. 2—Major R. D. Bennett, M.C. (Middlesex Regt.)
G.S.O. 3—Capt. J. F. Ferguson (Durham Light Infantry).
G.S.O. 3—Capt. E. R. Kennedy (Indian Army).

ADMINISTRATIVE AND QUARTERMASTER-GENERAL'S BRANCHES

Asst. Adjutant and Quartermaster-General—Colonel W. D. S. Brownrigg, D.S.O. (Sherwood Foresters).
Deputy Asst. Adjutant-General—Major J. A. Churchill, M.C. (Durham Light Infantry).
Deputy Asst. Quartermaster-General—Major M. J. A. Jourdier, D.S.O. (East Surreys).
Staff Captain—Capt. B. O. Hutchison (7th Hussars).

SERVICES

Officer Commanding Royal Army Service Corps—Lt.-Col. D. C. Cameron, O.B.E.
Asst. Director Medical Services—Colonel H. Ensor, C.B., C.M.G., D.S.O., M.B. (R.A.M.C.)
Asst. Director Ordnance Services—Colonel C. M. Stephen, C.M.G.
Senior Veterinary Officer—Major L. Daniels.

PERSONAL APPOINTMENTS

A.D.C.—Lieut. Knight (8th Hussars).
A.D.C.—Captain Stopford (Royal Artillery).

SPECIAL APPOINTMENTS

Base Commandant—Colonel (Colonel on the Staff) H. de D. Martelli, D.S.O. (R.A.)
D.A. Q.M.G.—Major W. W. Wagstaff (Beds. and Herts. Regt.).
Deputy Asst. Provost Marshal—Major J. A. Jervois, D.S.O., M.C. (K.O.Y.L.I.).
Senior Chaplain—Rev. M. W. Shewell.
Commanding Royal Artillery – Col. E. F Shewell, C.M.G., D.S.O.
Commanding Royal Engineers—Major E. A. H. James.

BRIGADE COMMANDERS

13th Brigade—Col.-Com. W. A. Blake, C.M.G., D.S.O.
14th Brigade—Col.-Com. Sir J. L. Burnett (Bart. of Leys) C.M.G., D.S.O. (Gordon Highlanders).
9th Indian Brigade—Col.-Com. P. B. Sangster, C.M.G., D.S.O.

May 1927

REGIMENTS AND UNITS AT PRESENT IN S

13th Infantry Brigade Headquarters.
14th Infantry Brigade Headquarters.
9th Indian Infantry Brigade Headquart

Royal Artillery :—
16th Medium Battery.
11th Field Battery.
80th Field Battery.
12th Pack Battery.
15th Pack Battery.
20th Pack Battery.
21st Pack Battery.

Royal Engineers :—
10th Field Company, Sappers and Mine

Royal Corps of Signals.

Infantry :—
2nd Battalion Coldstream Guards.
1st Battalion Devonshire Regiment.
2nd Battalion Suffolk Regiment.
1st Battalion Bedfordshire and Hertf ment.
2nd Battalion Gloucestershire Regiment
1st Battalion Border Regiment.
2nd Battalion Border Regiment.
2nd Battalion Durham Light Infantry.
1st Battalion Green Howards.
1st Battalion Middlesex Regiment.
12th Royal Marine Battalion.
4th Battalion 1st Punjabi Regiment.
3rd Battalion 14th Punjabi Regiment.

Royal Tank Corps :—
5th Armoured Car Company.

Royal Army Service Corps :—
No. 12 Mechanical Transport Company.

Royal Army Medical Corps :—
1st Indian General Hospital.
3rd British General Hospital.
33rd Field Ambulance.
12th Field Ambulance.
No. 3 Motor Ambulance Column.

Royal Army Ordnance Corps.

Royal Army Pay Corps.

Royal Air Force Squadron (No. 1).

REGIMENTS AND UNITS SHORTLY

The following regiments and units ar are shortly due :—

Sailed on the Transport Derbyshire on
2nd Battalion Welch Regiment
the Middlesex Regiment.

Sailed on the Transport City of Mar 10 :—

2nd Battalion Northamptonshire
2nd Battalion Scots Guards and a
the Royal Air Force.

Details of the Shanghai Defence Force, 1927.

...CE FORCES

...anghai, or to arrive here in the near future, is in the ...arly 20,000.

Sailed on the Transport Neuralia on April 20:—
Royal Air Force Squadron (No. 2).
Sailed on the Transport City of Poona on April 22:—
Two batteries of the Royal Field Artillery.
Sailed on the Transport Nevasa on April 23:—
1st Battalion Queen's Royal Regiment and the remainder of the 2nd Battalion Northamptonshire Regiment.

AMERICAN

The American forces (naval and marines) now in Shanghai total about 4,600 officers and men. Admiral C. S. Williams commands the naval forces, and Brigadier-General Smedley D. Butler is in command of the marines. The forces are made up as follows:—

1,600 naval officers and men under Admiral C. S. Williams.

1,500 officers and men of the Fourth Regiment of Marines under Col. C. S. Hill.

1,500 officers and men of the Sixth Regiment of Marines.

A U.S. Navy official statement announces that the American Marine Force now in China, or being concentrated at available ports in the United States for service in China, numbers approximately 9,000 officers and men.

FRENCH

French forces (including French Colonial Infantry and Annamites) now in Shanghai total about 3,500 officers and men. Admiral Desiré, Commander-in-Chief of the French Asiatic Fleet, is in command.

JAPANESE

Japanese forces immediately available in Shanghai total about 5,000. Some 2,000 have already been landed and are now engaged in patrol duty. Many thousands are available within 72 hours.

OTHER UNITS

In addition to the above there are also Italian, Portuguese, Spanish and Dutch forces actively engaged in supporting the Shanghai Defence Force.

MUNICIPAL DEFENCE FORCES

Commander—Colonel W. F. L. GORDON, C.M.G., D.S.O.

Shanghai Volunteer Corps— Total Strength about 2,000.

Units

Light Horse	"B" Company (British)
American Troop	American Company
Fd. Arty. Battery (Hows)	Portuguese Company
Light Gun Battery	Japanese Company
Engineer Company	Chinese Company
Machine Gun Company	Shanghai Scottish
Italian Company	Reserve Company
"A" Company	Russian Unit

Municipal Armed Police

Foreign
Foreign (Specials) } 694
Japanese
Sikhs 513

Maj General Sir Cecil M. Smith KBE CB MC. He was the Officer Commanding 12 MT Company RASC 1929 in Shanghai; he later became Director of Supplies & Transport, 1948-1951.

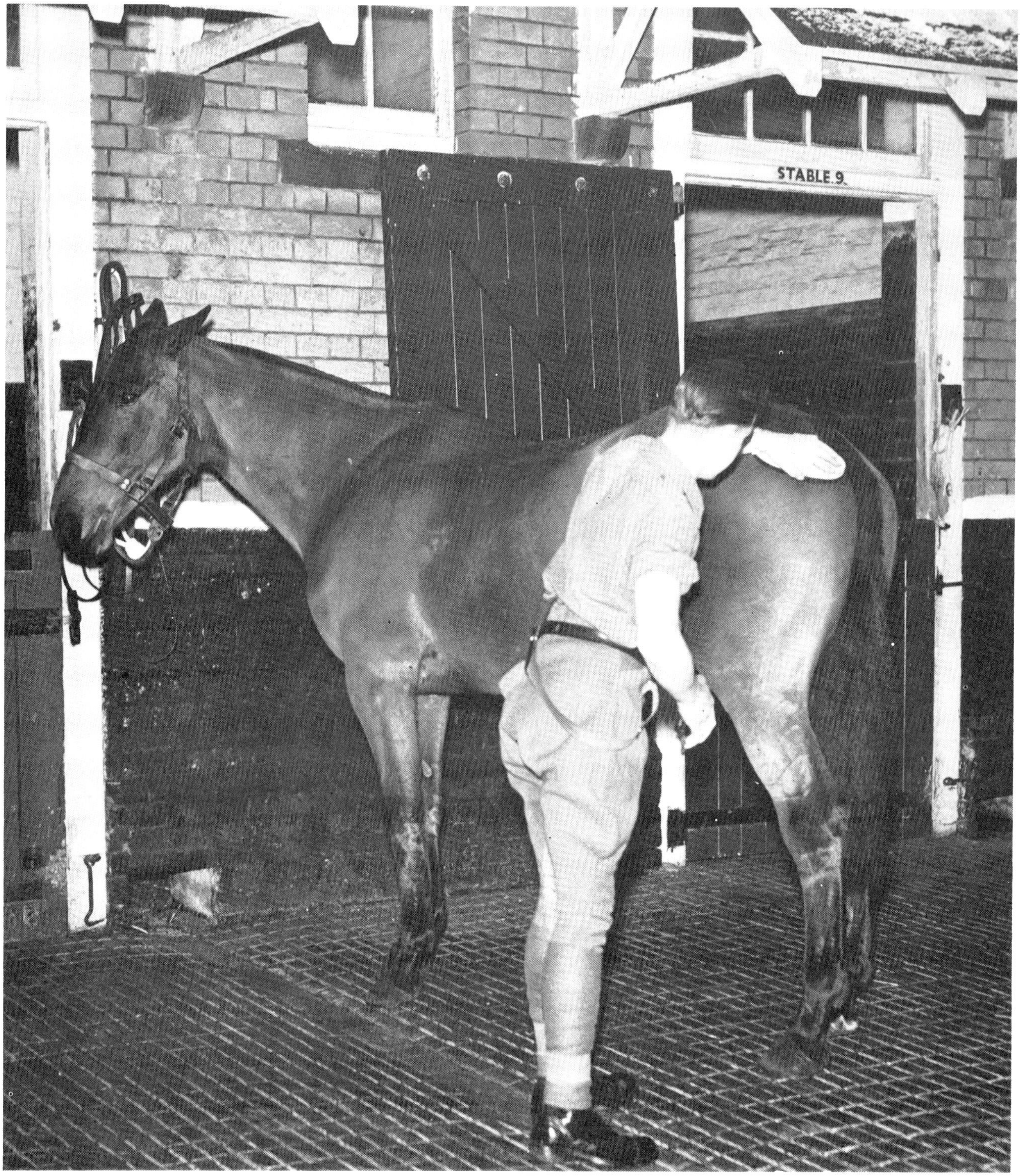

The RASC mounted soldier's first duty — the care of his horse.

REQUIEM FOR THE HORSE VIII

The reorganised Army Service Corps was formed at the end of 1888 and established as a combatant corps by Royal Warrant published on 2 January 1889, to carry out the dual functions of supply and transport; it inherited, from the Commissariat and Transport Corps, two horse transport depôt companies and thirty-four horse transport service companies.

By the time mechanical transport made its first appearance in 1903 (taking over the initials MT, which had previously been the familiar abbreviation of Military Train), there were seventy-three horsed service companies, of which ten were in South Africa, and three depôt and two remount companies. In 1911 a decision was taken to disband twenty horse transport companies and to form eight new mechanical transport companies. The writing was on the wall. The disbandment of horse transport and its replacement by mechanical transport had begun.

At the outbreak of the war in 1914 the number of mechanical transport companies had risen to twenty, and there were forty horse transport companies and two horse transport depôt companies.

The enormous expansion of the Army Service Corps during the 1914-18 war years involved the formation of 365 additional horse transport units, making a total of 405 companies, excluding the many other forms of animal transport raised. Most of these were included in the 78 divisional trains which took part in the campaigns in France and Flanders. Each train had about 400 heavy draught horses and a number of light draught horses and mules. The horse transport units operated mainly in the forward areas, backed up by mechanical transport in the rear areas, where suitable roads existed. The number of men serving in horse transport units reached a peak of nearly 58,000, including drivers, farriers, saddlers and wheelers. The horse had won a reprieve.

Yet the war also provided a great impetus to the development of mechanical transport and, although cross-country performance was still in its infancy and horse transport an essential requirement everywhere except on metalled roads, there were many enthusiasts who considered that it was out-dated, and that all the Corps' efforts should be bent towards rapid 'mechanicalisation' (as it was then called). There were heated arguments on the future of the Corps. Should it become a 'specialist' corps, the mechanical transport Corps of the Army, or should the regimental character be retained, with mechanical transport just one more of the special subjects in which the Corps should develop its expertise? It was the argument of the specialist versus the maid-of-all-work or the jack-of-all-trades. Those opposed to specialisation argued that the operation of mechanical transport was intrinsically the same as the operation of animal transport. As the latter required the specialists of the Veterinary Corps, so the operators of the former need not themselves be the specialists, as long as sufficient technicians were available. It was a good thing the enthusiasts were there, because they provided the stimulus which led to the early development of the 6–wheeled lorry and the pneumatic tyre. As Lieutenant Colonel G. le Q. Martel DSO MC RE wrote in his book *In the Wake of the Tank:* '. . . the development of the six-wheeled lorry was originally launched by a small band of enthusiasts in the RASC with little outside support or encouragement.'

In the end, the specialist lobby failed to win the day, and during the two decades following the end of the war, the trend gradually swung more towards the regimental view.

The end of the war brought a swift reduction in the number of transport units. By the end of 1920 and the beginning of 1921, when the size of the peace-time Army had been settled and the regular Royal Army Service Corps units reorganised and renumbered, there were only thirty-one horse transport companies left, nine fewer than in 1914.

Although the authorities were unwilling to give the *coup de grace* to horse transport, for it was still uncertain whether all its tasks could be satisfactorily performed by mechanical transport, and ten years were to pass before the final decision to replace horses with mechanical transport, the conversion and disbandment of horse transport units continued. During 1922 and by the beginning of 1923, twenty-three horse transport companies had been either amalgamated or disbanded; some of them were converted to mechanical transport. By 1924 there were only eight horse transport units left in the Royal Army Service Corps. At Woolwich, the original home and training centre for horse transport, there remained 'Y' (HT) Depôt Company – the RASC horse transport training company — and 3 (HT) Company, a service company. At Aldershot there were two service companies: 1 (HT) Company and 24 (HT) Company. 2 (HT) Company was at Bulford. 4 (HT) and 5 (HT) Companies were in the Rhine Army and 25 (HT) Company was in Eygpt. 24 (HT) Company had arrived in Aldershot from Northern Ireland in March 1923, after having had the distinction of providing a mounted escort for the Duke of Abercorn on his state entry into Belfast as Governor of Northern Ireland.

This reduction in horse transport was a sign of the times. Military thinking focussed on the concept of mobile forces, and collective training exercises were designed to test mobility and the administration of fast-moving mechanised units. Although it appeared certain that mechanical transport would eventually replace the horse, it was accepted that cross-country performance still needed improvement, and that in some theatres animal transport would be essential. It was agreed that a reserve horse transport company should be retained in the organisation of the Divisional Train. After the Aldershot Command collective training exercises held in September 1924, the special correspondent of *The Times* wrote, under the heading 'Triumph of Mechanical Transport':

'Supply and Transport services demonstrated in fact that the horse, so far as the Army is concerned, must soon become, if not extinct, then at all events a rarity. In an empire which extends over every type of country, it cannot disappear; nor with the increase of mechanical commercial traffic, can the training of horsemen be left to the chances of civil life. Pack batteries – mule or pony – will be needed so long as mountains and mountain paths and passes are likely to figure as obstacles to the progress of an army... Horses and horsemen will remain integral features of the Army, but there seems to be no reason why their mobility should not be increased enormously by mechanical means.'

Two years later it was decided, in October 1926, to withdraw 4 (HT) Company from the Rhine Army to Catterick and 5 (HT) Company soon followed, sixteen months later, in February 1928 and went to Aldershot. Six months after its arrival there 5 (HT) company was disbanded on 30 August. The company was not re-formed again until 16 October 1935, when it was revived as a mechanical transport company for service in Egypt.

The last horse transport company overseas, if we except units like 33 Company in Gibraltar, which kept some horse transport for local details, was 25 (HT) Company, which was converted to mechanical transport in Cairo in October 1928. Only a month later the disbandment of 3 (HT) Company took place at Woolwich and a reduction was made in the establishment of 'Y' (HT) Depôt Company. 3 Company claimed to be the oldest transport service company in the Corps.

A few months later 24 (HT) Company was disbanded and was not re-formed as a mechanical transport company until 1931, when it became part of the newly organised 1st Divisional Royal Army Service Corps at Aldershot.

This left only four of the eight horse transport companies which had still existed in March 1928, one at Woolwich ('Y' (HT) Depôt Company), one at Aldershot (1 (HT) Company), and one each at Catterick and Bulford (4 and 2 (HT) Companies).

Before horse transport could be finally replaced by mechanical transport there were two essential conditions on which the authorities had to be satisfied. First, the cross-country performance of mechanical transport should be at least equal to that of horse transport in a normal European theatre of operations. Secondly, sufficient mechanical transport of a suitable type must be available from civilian sources, to meet the Army's vastly expanded requirements on mobilisation. This latter problem was similar to the one which faced the Quartermaster General preparatory to 1914 with regard to remounts. An eight to ten fold expansion of the peace-time establishment of horses had been needed to meet immediate mobilisation requirements, and this large number had had to be found from civilian resources. A carefully prepared subsidy scheme had solved the problem in 1914. A similar scheme had been introduced in the early 1920s, to ensure that sufficient mechanical transport would be available from civilian sources, and Major F. L. Hewson was appointed to the important post of Chief Inspector of Subsidised Transport on 1 July 1926, and was succeeded four years later by Major E. H. Fitzherbert.

The introduction of the pneumatic tyre and the highly successful six-wheeled 30 cwt lorry (for which the research organisation of the RASC Training Centre at Aldershot could claim most of the credit) went a long way towards meeting the Army's requirements. By 1929 the horse had been largely replaced by the motor car and lorry in the civilian world, and it was realised that there would be difficulties in finding sufficient remounts in the event of mobilisation.

The 1928 manoeuvres were designed to test a new divisional organisation, in which there was no provision for the reserve horse transport company in the Divisional Train. A year later (in September 1929) a new RASC divisional organisation was introduced in place of the Divisional Train. The Horse Transport Company was omitted, completing the mechanisation of the divisional RASC.

This was the beginning of the end. Early in 1929 it was announced that two more horse transport companies would be disbanded. In May Woolwich, which for 75 years had been one of the chief Royal Army Service Corps stations and for over 30 years the home of horse transport training, bade farewell to 'Y' Depôt Company which, together with the regimental coach, moved to Aldershot, which now became the centre for all Royal Army Service Corps training.

In September 1929, 1 (HT) Company, the senior company in the Corps, paraded for the last time on the square at Aldershot, and in the middle of December, 2 (HT) Company disbanded at Bulford, maintaining to the last the high standards of smartness which had been a tradition in horse transport units.

The final curtain fell on the last day of the year — 31 December 1929 — when 4 (HT) Company, the last of the horse transport service companies, was disbanded at Catterick. The company had previously been 296 (HT) Company, when it formed part of the Rhine Army Train, and had marched from France to Cologne as one of the original units of the army of occupation. It was renamed 4 (HT) Company in 1924 and spent two more years in Germany. During its three years at Catterick it competed in many mounted events and local shows.

Among the many prizes won were two firsts and a second for GS Wagon and Pair, two firsts in Individual Turn-outs and various others for jumping and tent pegging. To the end they kept up the best of the old spirit and tradition of horse transport.

With the end of 1929 the horse transport service companies passed into history. But their spirit and tradition did not die. At the last inspection of the last company in the last year of their existence, the Assistant Director of Remounts wrote to the Commander in Chief of Northern Command: 'One of the best turnouts I have ever seen, an object lesson to any unit.'

With the disbandment of the last of the horse transport service companies, 'Y' (HT) Depôt Company, shortly to be renamed HT Training Company, Animal (Transport) Training Company and finally Transport (Animal) Company, was left as the only horse transport unit in the British Army. Henceforth its role was to be the training of Royal Army Service Corps officers and men in the technique of animal transport, in case this form of transport might be required in a future theatre of war. For this purpose an establishment of two officers, four warrant officers, 34 NCOs and 65 men only were allowed. There were not many who could foresee that more than 30,000 mules would be required in North Africa and Italy – a European theatre of operations – in the 1939/45 war.

The military correspondent of *The Times* devoted a full column and a half to a moving tribute to the part played by the horse transport of the Royal Army Service Corps. The article ended with these words: 'The familiar GS Wagon and "limber" will not entirely disappear at present, though thousands must have been scrapped since the War. The first line transport of all arms is being gradually mechanised, but many infantry battalions and artillery brigades retain horsed vehicles in their regimental transport, pending the result of experiments with those units that have been completely mechanised. A few horsed wagons still remain with the divisional trains and ambulances of the Territorial Army, but it may be assumed that their days are numbered. Soon the Royal Army Service Corps will have nothing to remind it of these obsolescent vehicles except its regimental march, "Wait for the Wagon" '.

In 1932 horse transport was withdrawn from the divisional trains of the Territorial Army. The virtual end of horse transport in the Corps aroused mixed feelings. For most there was a certain nostalgia at the loss of what amounted to a way of life. The following extract from an editorial in the *RASC Journal* of January 1930 sums up contemporary feelings: 'For many reasons, the passing of the old HT Companies will be deeply regretted in the Corps. It means the loss of a certain humanising influence, as no one can fail to appreciate who has known the devotion of the transport driver to his animals, even in the most trying circumstances.'

The 'humanising' influence of the horse has more than once been remarked on. The precept that an officer's duty lay first in the care of his horses, then of his men and finally of himself, was reflected in the general attitude of horsed units towards discipline and smartness in turnout. The horse had romance, an epithet which could hardly be applied to an Army lorry, although some of the early enthusiasts did their best to pretend that it could. Animal management demanded a strict time-table. Regular feeding and watering were essential; the care of an animal could not be put off, as the cleaning of a carburetter or the filling of a petrol tank could be.

The Corps set high standards in the turnout of its horse transport units and these were maintained to the end. Final inspections before disbandment of horse transport units received glowing tributes from senior regimental and veterinary officers.

The reputation of the Royal Army Service Corps in mounted events stood high. From Woolwich, Captain J. R. Allen staged an impressive team jumping event at the Royal Tournament at Olympia every year, until he was posted to Bere Island in 1928. And he, with Major Pereira, Captain Kerr, Lieutenant-Colonel O'Hara and Lieutenant Somerset, had many show jumping successes. The Corps coach was always prominent among military coaches and successfully challenged many of the leading civilian coaches, which had unlimited financial backing, at the various horse shows. Captain H. R. Kerr was elected a member of the Coaching Club in 1923, an honour at that time shared by only two other Army members, both gunner officers. He was joined as a member in 1930 by Lieutenant Somerset who, as riding instructor at Woolwich and later at Aldershot, upheld the reputation of the Royal Army Service Corps in show jumping and other mounted events at horse shows and race meetings. Just before its disbandment, 2 (HT) Company, at Bulford, won first prize in the turn-out class for GS wagons, out of fourteen entries from all arms in the command.

Until the 1960s the Royal Army Service Corps continued to achieve successes in mounted events at the Royal Tournament and in command competitions. In the spring of 1932 a team from the Transport (Animal) Company of the Training Battalion won the team jumping competition in the Bronze Medal Tournament, which was open to all units in the Aldershot Command, in the face of strong opposition from Cavalry, Royal Artillery, Royal Engineers and other horsed units. Members of the winning team were: Sergeant Reedman on Gem, Sergeant Webb on Gambler, Lance Corporal Degnan on Ruby and Lance Corporal Johnston on Gloaming.

In 1963 Major G. Boon, who was the third member of the Corps to be elected a member of the Coaching Club, staged a popular team jumping event at the Royal Tournament at Earls Court, and at the Windsor Horse Show the Corps team won the Queen's Cup three years running. The year before Major T. J. 'Topper' Brown added to his many successes by winning the King's Cup at the Royal Tournament on 'Finisterre'.

When the last horse transport service company had been disbanded, the Corps enjoyed some of its major successes in the mounted field, with the two coaches at Aldershot, Colonel O'Hara at the races in Malta, Major C. N. Custance and Lieutenant E. J. J. Pearn in Gibraltar, and the continued success of the Corps point-to-point at Shipton Bellinger near Bulford, largely due to the enthusiasm of Lieutenant Colonel A. M. Wilson.

For many years horse transport and mechanical transport had marched side by side, complementing each other. Many officers who, in the early days, had been in the forefront of mechanical transport development before 1914, were among the keenest horsemen between the wars. Colonels O'Hara and Foster and Lieutenant Colonels Gill and Wilson were outstanding examples of this shared enthusiasm

Right up to the outbreak of the war in 1939 all officers joining the Royal Army Service Corps between the wars underwent a course in riding, animal management and pack transport, first at Woolwich and then, when 'Y' Depot Company moved to Aldershot, at the RASC Training Centre.

It was fortunate that, when the Corps went to war again in 1939, there were still many left in the Royal Army Service Corps and the Royal Indian Army Service Corps who had served in horse transport units and understood the operation of animal transport. Without them and their experience it would have been a far more difficult task to raise and train the many mule companies, which played such an important part in keeping troops supplied in mountainous areas.

ENVOI: 'No village maid, returning from the Fair,
Waits for the helpful waggon on its way.
Mechanic monsters earth and air pervade.
Good transport horse, long-faithful friend,
Farewell!'

Lord Darling
(Lord Chief Justice)

Note: Charles John Darling, 1st Baron, c1924, of Langham. Born 6 December 1849. Judge of the King's Bench Division of the High Court of Justice 1897-1923. Died 29 March 1936. Among publications: *A Pensioner's Garden* (1926), *Reconsidered Runes* (1930), *Autumnal Leaves* (1933).

ABOVE: A horse's task is carrying loads. The soldier had to have a thorough knowledge of the weight distribution and harness needed for individual loads. BELOW: Two well-known old faithfuls: on the right is Lorna Doone; the stable companion, Kingsley.

Toothache.

ABOVE: Tandem jumping and BELOW: tandem driving.

A delicate moment.

1 (HT) Company RASC on parade, Aldershot, 1929.

Friends — outside the Riding School.

PART II
1930 TO 1939

THE TURN OF THE TIDE IX

'It is the hour, not for despair, but for courage and rebuilding; and that is the spirit which should rule us in this hour.' Sir Winston Churchill, 4 October 1938.

When Major General Sir Evan Gibb KBE CB CMG DSO became Director of Supplies and Transport in August 1929, prospects for a settled period were reasonably bright. Ten years had passed with no real threat to world peace. Most trouble spots seemed comparatively quiet. Mussolini had cleared up many corrupt practices in Italy and had brought a degree of prosperity to his country. Only the benefits of dictatorship were so far outwardly apparent. In Germany the anti-Communist elements seemed to be getting the upper hand. Although, in both countries, authoritarian influences were making headway, neither appeared to have the strength to threaten the peace. Although the League of Nations had not achieved all the optimists expected, machinery had been created for peaceful settlement of international disputes. If only limited progress had been made towards disarmament, talks continued. Politicians promoted disarmament and believed full–scale armed conflict could be avoided. 'No major war for ten years' was still the planning assumption. There was no sign of the imminent collapse of the American economy.

This period of euphoria was short-lived. Less than three months later came the Wall Street crash, revealed in all its immensity on 29 October 1929, a day to be remembered in the United States as 'Black Tuesday'.

Although the effects on the British economy were not immediate, they were sufficient to bring the second Labour Party administration to an abrupt end in less than two years, leading to the establishment of a National Government under Ramsay MacDonald on 26 August 1931, which was compelled within less than a month to abandon the gold standard — on 21 September 1931. The crisis of 1931 was followed by recession, in which unemployment rose alarmingly and the economy suffered severely. Drastic cuts were made in salaries and unemployment benefits; economies were made wherever possible. Inevitably some fell on the Services. Expenditure on the mechanisation programme was postponed; large-scale training exercises were cancelled and Territorial Army training reduced: establishments were pruned of their non–essentials and cuts were made in rates of pay. The Duke of Connaught gave up his annual visit to the Riviera, and stayed instead at Sidmouth. Various Corps functions, including the point-to-point and the Corps Club tea-party and dinner, were cancelled.

The depression did not have as great an impact on the Royal Army Service Corps as might have been expected. Owing to the recent disbandment of the last of the horse transport service units, the Corps was not asked to make any further reductions in establishments. In any case, there was little fat to spare. The Corps could barely meet its peace-time commitments. There were, of course, reductions in expenditure on development projects, and collective training was curtailed. But in other ways the Corps benefitted. As always, in times of high unemployment, recruitment showed a sharp improvement, both in numbers and in quality. The end of the horse transport era allowed the Corps to concentrate on its mechanical role. The reorganisation of the Training College and the transfer of responsibility for Royal Army Service Corps training to the General Staff gave a new impetus to the serious study of training for war and a greater sense of purpose was apparent. The tone of the *Journal* reflected this in an editorial published in September 1930, under the heading: 'A Spot of Optimism'. The writer pointed out that, in spite of the depressing stories in the Press of unemployment, lack of trade, the depreciating pound, undiscovered murders, more taxation, and general gloom, there was no cause for pessimism in the Corps. Our cricketers had just beaten the Royal Artillery, the Royal Corps of Signals, the Royal Tank Corps and the Royal Army Ordnance Corps; our Catterick tennis pair had reached the finals of the Army Championships; at Aldershot we had won the command water polo and team swimming race; members of the Corps had represented England at fencing and had figured in the Army and English hockey elevens. The regimental coach had taken on all military rivals, and a good few civilian competitors, and beaten them. And in the more serious field of soldiering, we had entirely reorganised our training for war and had established regular war courses at the Royal Army Service Corps Training College; our divisional war organisation had been put on a sound footing and a firm regimental basis. Two officers were due to go to the Staff College in the following January. Our recruits were better than ever. The new *Quarterly* had been launched and had been enthusiastically received by the Army. These were hardly reasons for the Corps to be downhearted.

But the first decade came to an end with the Army still unsure of itself. The 'Ten-year rule' was still in force, renewed in 1929 after a review when Churchill was Chancellor of the Exchequer. This was eagerly seized on by the Treasury to justify imposing economy on the Defence Vote and had a stultifying effect on any attempt to develop a sound and realistic strategy.

The role of the Army was still not clear. There were two schools of thought; one, that we should concentrate all our efforts on the defence of the homeland and the protection of our vital overseas interests; the second that involvement on the Continent of Europe could not be avoided, and our preparations must include a continental expeditionary force.

The threat of massive air attack was a new and disturbing factor. The development of modern air forces with their ever increasing power of damage made air defence one of the main pillars of defence strategy. The role of the defence forces seemed to be: (i) to provide adequate air defence; (ii) to secure our overseas possessions, particularly in the Middle East and India, where a large part of the British Army was deployed. This involved the protection of our communications and the territories where our vital interests lay; (iii) to defend the United Kingdom; (iv) possibly to provide an expeditionary force for the Continent.

During the previous ten years the Corps had settled down into a peace-time routine, meeting day-to-day administrative commitments, playing its part in dealing with sporadic outbreaks of unrest, and carrying out what training other tasks permitted. The Army and the Corps had relied, for an extension of this minimal role, on the reserves of trained manpower and the experience of those who had served in the last war. By now, almost a third of the officers and the majority of other ranks no longer had any active experience. The time had come for reorientation and readjustment.

The Corps' commitments in India ended. The Rhine Army had come home. The troubles in the Far East had subsided. The recession forced economies and sharpened the test of what was really necessary. Disarmament was proving illusory. The failure of the Disarmament Conference in Geneva in 1932 was a disappointment and destroyed many people's faith in the

League of Nations. The growth of the National Socialist Party in Germany cast a shadow over Central Europe. Disturbances in Palestine, as the number of immigrant Jews approached parity with the Palestinian Arabs, underlined the growing influence of that area on Imperial strategy, and the importance of Egypt as a military base.

Horse transport had at last given way to mechanisation; the Corps was committed to an organisation in which the horse no longer played any part, except in the nostalgic retention of a diminishing technique which nevertheless proved of great importance in the next war.

Development in mechanical transport and in armoured fighting vehicles since the war gave clear signs that a future major conflict would be quite different. Highly mobile forces supported by a greatly improved air component would alter the whole character of warfare and would prevent a repetition of the static trench warfare of the Western Front. The main emphasis in major training exercises was on mobility, traffic control, air defence and the development of an administrative organisation, able to support rapid advances over long distances. In this concept the Royal Army Service Corps had a major role to play.

While the recession lasted, no expansion in the size of the Army or the Corps could be expected, nor any substantial measure of rearmament. But the failure of the Disarmament Conference in 1931, followed by Germany's withdrawal in 1933, sounded alarm bells. In 1934 a Defence Requirements Committee reported to the British Cabinet early that year; the Chiefs of Staff had assessed that Germany might be ready for war by 1938 or 1939, and recommended a five-year rearmament plan. It was at last admitted, albeit reluctantly, that war was a distinct possibility and the Government decided to abandon the 'Ten year rule' and to embark on a programme of rearmament. The emphasis shifted sharply towards involvement on the Continent of Europe and the urgent need for anti-aircraft defence and armoured formations. Only a small part of the necessary finance for such a costly programme was initially forthcoming. Meanwhile, the most that the Corps could do was to put its house in order so that when the word was given, advantage could be taken of this dramatic change of policy.

Although the Royal Army Service Corps in 1930 was smaller than it had ever been since 1914, or than it would be after 1939, it was compact, well-balanced and professionally well-qualified. There were substantially fewer than 500 officers, including quartermasters and mechanist officers. Although there was only one major general's appointment — when Colonel H. N. Foster was promoted Major General from being Assistant Director of Supplies and Transport, Southern Command, there was no post for him to occupy and he had to remain on half-pay, until he retired—the number of colonel's and lieutenant colonels' appointments provided a healthy rank structure, with good career prospects up to those ranks. The Royal Army Service Corps held a good proportion of staff appointments and there was a small but regular flow of students to the Staff College. There was a strong 'family' feeling in the Corps, which in size was comparable to the Royal Marines, which was perhaps not so noticeable in larger Corps like the Royal Artillery or the Royal Engineers.

Reorganisation of RASC training, both in War Office responsibility and in the Training College at Aldershot, had opened the way for a vigorous and enlightened approach. Although there was no significant expansion of the Corps before the middle 1930s, a number of important changes in organisation took place which had far-reaching effects. Foremost was the creation of divisional RASC companies as an integral part of the formation which they served. The formation of separate petrol companies foreshadowed the development of a whole system of petroleum supply, which so far had only been treated as an academic study and hardly put to the test on any major exercise. Later, the double echelon supply system, which had been used in the 1914-18 war for cavalry formations, was introduced as an integral part of the maintenance system for supplying fast-moving mobile forces. The rapid growth of National Socialism in Germany gave an added impetus to all these developments.

Between 30 June 1930, when the occupying troops left the Rhineland, and the beginning of 1933, the National Socialist Party made great strides in Germany. The ageing President, von Hindenburg, was at last persuaded that the only cure for Germany's ills and hope for the future lay with the National Socialists and he gave his support, which carried with it the influence of the old German army, to the 'Fuhrer' of the party. On 30 January 1933 Adolf Hitler was sworn in as Chancellor of the Reich by Hindenburg.

From this day the signs were clear to all those who wished to see them. The strident voice of the Nazis proclaimed their intentions and one act after another made it clear that sooner or later war was inevitable.

Hitler's success in Germany, which had aroused only half-hearted reactions in France and Great Britain, gave encouragement to Mussolini, who launched his Abyssinian campaign in 1935. When the Spanish Civil War broke out on 18 July 1936, both dictators saw the opportunity to put their military training and equipment to the test. By the end of the war on 28 March 1939, many valuable lessons had been learned.

By now the composition of the British Expeditionary Force had been determined. The leading echelon was to be one Army Corps of two divisions (1st and 2nd Infantry Divisions) based on Aldershot Command, followed by a second echelon consisting of a second Army Corps, also of two divisions (3rd and 4th Infantry Divisions) based on Southern Command.

Meanwhile Hitler had greatly strengthened his position. The outcome of the Saar Plebiscite in 1935 tempted him to denounce the Treaty of Versailles on 16 March and to declare that Germany's air force had reached parity with the Royal Air Force. The invasion of Austria in March 1938, followed by the incorporation of Austria in the German Reich, caused widespread alarm, which was intensified by a similar threat to Czechoslovakia a few months later. Diplomatic moves, culminating in Neville Chamberlain's flights for discussions with Hitler, only served to postpone the crisis and to give the Allies a short respite in which to hasten rearmament.

The invasion of Czechoslovakia and the destruction of its republic on 15 March 1939, only twenty years after its creation, in violation of the four power agreement between Britain, France, Italy and Germany, signed in Munich on 30 September 1938, gave a final unequivocal warning, and preparations for the inevitable war accelerated.

The way ahead was clear. It was no longer sufficient for the Corps to do its best to meet its various commitments, as they came along, with whatever resources were available. It had to gear itself to meet the demands of war. This involved serious study of likely problems, the creation of a relevant organisation, and the expansion of resources.

As the political situation deteriorated and the prospect of war became more certain, urgent steps were taken to expand the Corps by the creation of many new units, by stepping up recruitment of both officers and men and, eventually, by the introduction of military service and a new militia. At the same time urgent preparations were made to increase reserves of supplies and petroleum both at home and abroad, particularly in the Middle East, and to put the scheme for subsidised transport on a war footing. A prodigious amount of work was undertaken by the War Office staff of QMG 3 and QMG 6 during the months before the outbreak of war.

By the summer of 1939 it was clear that it was too late to buy more time for rearmament and, when the threat against Poland materialised, although the possibility of direct intervention was remote, the ultimatum, which was to plunge Europe once more into war, was at last delivered to a ruthless and unrelenting tyrant.

Major General Sir Evan Gibb KBE CB CMG DSO, Director of Supplies & Transport 1929-1933.

Major General M. S. Brander CB OBE MIMechE, Director of Supplies & Transport, 1937-1941.

THE WAY AHEAD X

Although His Royal Highness The Duke of Connaught celebrated his 80th birthday on 1 May 1930, he continued to take an active interest in the affairs of the Corps. In 1929 he had presided at the Corps Club Dinner. On 7 September 1932, in his 30th year as Colonel, he paid a comprehensive visit to Royal Army Service Corps units in Aldershot. Later in the year, on 3 December 1932, he was appointed Colonel in Chief of the Royal Army Service Corps.

In April 1937, while he was staying at Bournemouth, he made a point of visiting and meeting the members of 27 War Course of the Royal Army Service Corps Training Centre, who were involved in an exercise at Poole.

On the occasion of Golden Jubilee of the birthday of the Corps, on Sunday 11 December 1938, the following telegram from HRH The Duke of Connaught, Colonel in Chief the Royal Army Service Corps, was received by the Representative Colonel Commandant, Sir Evan Gibb:

> 'I send you my warmest congratulations on the splendid service rendered by the Corps during the past fifty arduous years, and I congratulate all ranks on their fine record'.
>
> (signed) *Arthur*, Colonel-in-Chief

Just before the outbreak of war, on 31 August 1939, in his 90th year, he visited the newly formed Militia Training Battalion RASC in Boyce Barracks, Ewshott, near Crookham, and was photographed with the officers of the battalion.

The three original Colonels Commandant, whose appointments in 1921 were for ten years or until they reached the age of 70, were succeeded by General Hobbs, who replaced General Clayton in 1925, General Ford, who succeeded General Landon in 1930, and General Gibb, who took General Carter's place, when he died in 1933. Generals Hobbs and Ford, in their turn, were followed by Generals Tarver and Macleod in 1935 and 1938. On the outbreak of war, General Gibb was the senior Colonel Commandant and General Macleod was the Representative Colonel Commandant for 1939.

During this second decade of the period covered by this history the affairs of the Royal Army Service Corps came under the direction of three officers of great ability and wide experience who, however, had very different personalities and who brought quite distinct qualities to the management of the Corps.

The first of these was Major General Sir Evan Gibb KBE CB CMG DSO, who first saw active service on the west coast of Africa in 1898 and in the Boer War, and subsequently served on the staff of the War Office. He then became an instructor at the Army Service Corps Training College, Aldershot, until the outbreak of war in 1914. During the war his organisational ability was directed to the control of labour in the BEF (as a brigadier general) and later as Controller of Salvage. For six months in 1920 he was Commander of the Forces in France and Flanders. As Assistant Director of Supplies and Transport, Aldershot Command ('temporary Colonel Commandant', the equivalent of Brigadier) in the middle 1920s, he was in close touch with the RASC Training College and with the collective training exercises held in the Command.

He was promoted major general in 1927 without any specific appointment and spent two years in association with the NAAFI, mainly in connection with overseas commitments. He was appointed Director of Supplies and Transport on 9 August 1929, only a few months before Colonel D. C. Cameron was due to complete his tenure as Commandant of the Training College and to join him in the War Office as the first officer to hold the dual appointment of AAG AG8/ADST QMG5.

General Gibb was a great extrovert and had a bluff and genial personality. He had a wide knowledge of administration, and combined his experience with a down-to-earth approach. He was quick to appreciate the value of Colonel Cameron's proposals for the reorganisation of Royal Army Service Corps training, and gave him his full support and the weight of his considerable influence in high places. The Quartermaster General then was Lieutenant General Sir Hastings Anderson KCB, who had been Commandant of the Staff College and who came to know the Corps when he was DQMG of the Army in India. General Gibb found him most receptive of new ideas affecting the organisation and training of the Corps which lost a very good friend at his early death in December 1930.

The partnership of General Gibb and Colonel Cameron brought about two of the most important post war decisions affecting the future of the Royal Army Service Corps. The first of these was the agreement that training for war of the RASC would in future be dealt with by the General Staff, and that supervision of training at the RASC Training College would be the responsibility of the Directorate of Military Training. Secondly, it was decided that all regular divisions of the Army should be provided in peace with their appropriate Royal Army Service Corps units. This meant that each of the four divisions would have its own RASC peacetime cadres which, on mobilisation, would form the framework for expansion to war strength. This was of the greatest importance, not only to the Royal Army Service Corps but to the whole Army, which in war was so dependent on the efficient performance of the services of supply and transport.

Before these decisions were taken there had been no central guidance or common doctrine on unit training. Procedures varied from command to command and the system of cost accounting of the employment of drivers and artificers had had a stultifying effect on training. When the responsibility was transferred to the Director of Military Training, the RASC began to be treated like the rest of the Army. Training was given first priority and peace administration was relegated to its correct sphere. As a result of the closer co-operation with other arms the rest of the Army learnt more about the RASC, its organisation and its methods of operation.

An important effect of this change of attitude was that the Royal Army Service Corps was entrusted with the carriage and delivery of all natures of ammunition, including artillery ammunition, previously the responsibility of the Royal Artillery.

The creation of the new branch AG8 brought the Corps into line with the rest of the Army with regard to postings and appointments. Henceforth they became the responsibility of the Adjutant General instead of the Quartermaster General.

The divisional ammunition, baggage and supply companies now came under the command of the GOC and were the sole responsibility of him and his staff. They were at his disposal for any exercise which he decided should take place, and they would mobilise with the rest of the division.

Before this, units had to be produced on an *ad hoc* basis. There was no opportunity for staffs to practise in peace the supply and transport requirements of their formations with the units which would go with them to war. There had been no chance of building up confidence between the fighting troops and the administrative services. It was fortunate for the Corps that the General Officer Commanding the 1st Division at Aldershot was an officer who had served for eleven years in the RASC. General Harman (Lieutenant General Sir A. E.

Wentworth Harman KCB DSO) transferred to the ASC from the 3rd Dragoon Guards in 1896 and remained with the Corps until 1908, when he rejoined his old regiment as adjutant. His last Corps appointment was Deputy Assistant Director of Supplies and Transport in South Africa. He never lost touch, expected the highest standard from the Corps units under his command, understood their problems and always gave them his support. While he was at Aldershot, many of his exercises centred around the handling of transport in the division and the problems of maintaining mobile formations. One of these (in 1931) involved equipping the division at full War Establishment in order to test the divisional organisation, and in another, given the title of 'The Aldershot Experiment', as many 30-cwt vehicles as possible were replaced with 3-tonners, in order to reduce the number of vehicles required for the maintenance of the division. He paid his old Corps the compliment of joining in the march past of ex-members in mufti at the Sunday parade in Corps Week, and he remained a member of the Corps Club until his death.

The change in responsibility for RASC training led to a complete reorganisation of the RASC Training College. Before 1929 the College existed mainly as an initial training ground for young officers and recruits. It provided no resources for senior Corps officers to study the more advanced war problems of their profession. Once an officer had completed his junior officers' course there were virtually no facilities for him to study progressively, away from the constant pressure of day-to-day administration, the organisation of the RASC in the field, or the many problems of supply, embracing such subjects as forecasts, provision and local purchase. For the officer above the rank of subaltern there was little else available, apart from courses at the Royal Sanitary Institute and the London School of Economics, for advanced study.

Now, four war courses a year were instituted with the object of studying every stage of administration from the base to railhead and beyond, during all the varying phases of mobile warfare. These courses were open, not only to Corps officers, but to officers of all arms, affiliated Corps and the other Services. At the time they were the only administrative training courses of their kind and provided an excellent preparation for the Staff College. The first of these courses ended in January 1930 and the final day was attended by the Quartermaster General, the Commandant of the Staff College and the DST.

An advanced supply course was also introduced with training for war as the theme, instead of the previous emphasis on 'Supply, Transport and Barracks Services'.

The interior organisation of the Training College, renamed in November 1931 'The Training Centre', was also changed radically, with the formation of a training battalion. The training company consisted of recruit squads which were given the names of distinguished Corps officers, which had an immediate effect on *esprit de corps* among young recruits.

Regular courses were also organised for NCOs, covering weapon training, anti-gas training, physical training and drill, which were open to all home stations, the object being to ensure uniformity in instruction and a free exchange of ideas between Aldershot and other stations.

The *RASC Quarterly*, which had been allowed to lapse in 1928, was given a new lease of life with a new series in 1929, which soon came to the fore as a serious military publication. This series differed from previous issues in that all the articles appeared anonymously, and were designed to express the balanced views of the Training Centre on current Corps studies and progress in the training for war. A new issue of training manuals was made 'by command of the Army Council', which contained authoritative data on which unit commanders could base their training programmes.

While General Gibb was at the War Office, the appointment of Director of Supplies and Transport was raised from that of a Grade II to a Grade I Director, in May 1933. Other landmarks were: the establishment of Catterick as a headquarters station ('the Aldershot of the North'); the reorganisation of the 14 Territorial Divisional Trains RASC, so that their units would conform to their regular counterparts; the modernisation of the Army School of Cookery (still one of the units of the RASC Training Centre), with a greater emphasis on cookery in the field under wartime conditions.

Although the economic situation which prevailed throughout General Gibb's appointment as Director hardly permitted any expansion of the Army, and the international state of affairs did not yet call for immediate rearmament, nevertheless the programme of mechanisation was allowed to go forward. The disbandment of the horse transport units had provided some scope for a moderate increase in the number of mechanical transport units. At the end of 1928 the total number of Royal Army Service Corps companies reached its lowest ebb, with only twenty-three companies at home, and ten companies overseas (including the local companies in Malta and Gibraltar). In October 1931 approval was given for the formation of four additional companies (24, 40, 41 and 42), in order to implement the decision to form divisional RASC units. Eighteen months later seven more companies (numbered 43 to 49) were authorised with effect from 1 April 1933. During the four years since General Gibb's appointment in 1929, not only had a most important reorganisation of the Corps taken place, but the number of RASC MT companies had been increased by more than 30 per cent. The tide was beginning to turn.

In an article in the *RASC Journal* of January 1934 the following tribute to General Gibb's stewardship appeared:

'There are two officers, late of the Corps and now retired, who were dominating figures in obtaining for the Corps the results we have endeavoured to record. This generation of the Corps acknowledges a great and sincere debt of gratitude to Major General Evan Gibb and Colonel Donald Campbell. It is to these officers, working in the closest harmony, that much is due. Their foresight, sagacity and determination have won the Corps a new and refreshing birthright. We can only express our appreciation and our admiration for their example'.

On 9 August 1933 General Gibb completed his four years as Director of Supplies and Transport. On 3 February he had been appointed Colonel Commandant RASC in succession to the late Major General Sir Evan Carter, an appointment which he held until his death in 1947. In 1940 he left his retirement to resume an appointment similar to the one which he held in the BEF in 1919, that of Director of Salvage in the War Office.

General Gibb was succeeded by Major General C. W. Macleod CB CMG DSO who had just completed his appointment as Inspector of RASC Services. General Macleod had a different personality. He had all the qualities of a high grade staff officer: great capacity for hard work, remarkable powers of concentration, a thorough grasp of detail, sound judgment and complete reliability. His unshakeable loyalty and innate modesty endeared him to his colleagues in the War Office and inspired the devotion of his own staff. Because of his great knowledge of staff duties and establishments, accumulated from his experience in the War Office in several capacities, his advice was always in great demand. He was never happier than when he was working with figures. During the war he had been Deputy Director of Transport under Sir Bertram Boyce in France, after which he was Assistant Director of Transport in the War Office for four years, followed by the appointment of Assistant Director of Supplies and Transport, Eastern Command, before returning to the War Office as Inspector. The keynote of his tenure as Director was expansion and preparation for mobilisation. The first important steps in this direction had been taken by his predecessor. From the date of General Macleod's appointment, conditions in Europe rapidly deteriorated.

The need to send reinforcements to the Middle East to protect vital interests made it clear how slender our resources were. More than fifteen years had passed since the end of the war; the country could no longer depend on the large body of reservists available at the Armistice.

During the four years of General Macleod's time as Director there was a gradual increase in the size of the Corps and further new companies were formed. In 1934 there were 467 Regular officers and some 5,000 rank and file. The Army estimates for 1936 allowed an establishment of 472 officers and 6,076 rank and file. In the following year these figures were increased to 508 and 7,162. There were thirty-four mechanical transport companies at the end of 1933. When General Macleod retired an additional seven companies had been formed. It is true that these companies were little more than cadre units, with only fifteen load-carrying vehicles, but they provided the framework for expansion when they were required to take on an operational role, as was successfully demonstrated when 'Saarforce' was constituted, and during the emergencies in Egypt and Palestine. Other measures were the doubling of the size of the Training Battalion at Aldershot by the addition of a second training company on 1 January 1936, the re-employment of retired officers in home stations to relieve regular serving officers for service overseas, and the enlargement of the organisation of the Chief Inspector of Subsidised Transport. Including the Chief Inspector (Major R. J. O. Dowse) there were now thirty-five inspectors.

The first of these measures had become necessary to deal with the training of additional recruit drivers, needed to man new units. In March 1937 there was a record number of recruits at Aldershot.

The woeful shortage of military transport vehicles became most apparent during the large scale collective training exercises held in 1935 and 1936 — the first since the restrictions imposed by the recession — when a considerable number of hired transport vehicles had to be used in order to bring units only partially up to strength. This forcibly underlined the importance of the subsidy scheme.

The Royal Army Service Corps played its part in the abdication crisis in December 1936 and the coronation of HM King George VI. Seven RASC motor-cyclists were on continuous 'Palace Duties' between 5 and 11 December 1936 to maintain communications between No 10 Downing Street and Fort Belvedere. In May 1937 the coronation placed a heavy administrative commitment on the Corps which was carried out with great efficiency.

General Macleod continued to give active encouragement to RASC officers to sit for the Staff College examination. This policy was already paying good dividends. At the end of 1933 no less than six senior staff appointments in the rank of brigadier and colonel in home and overseas commands were held by officers of the Corps. Two or three RASC officers were securing places on most Staff College courses, and others were qualifying without obtaining vacancies.

In presenting the Army Estimates in the spring of 1936, the Secretary of State announced improved conditions for the Army and defined its roles under four headings: (i) the protection of our naval bases; (ii) the policing of the Empire; (iii) the defence of our shores, and (iv) in emergency, to provide an expeditionary force.

This was the first time since 1918 that the provision of an expeditionary force for overseas service had formed part of the national defence strategy. Work could now begin on plans for raising the necessary units.

When General Macleod retired on 14 June 1937 he handed over to Major General M. S. Brander CB OBE MIMechE the direction of a Corps which was already aware of the likelihood of war, which was beginning to expand, and which had already started to study with great seriousness the problems which the future was likely to present. Although the Royal Army Service Corps was still a small Corps, it was compact and well-balanced, with a growing proportion of staff trained officers and an overall professional outlook.

On his retirement General Macleod immediately joined the staff of the Director General of Munitions Production (who was a member of the Army Council) and was appointed Area Progress Officer for Scotland, with his Headquarters at Glasgow. At the beginning of the war, with great loyalty, he returned to the War Office as a retired officer, to serve as Deputy Director of Supplies and Transport under his successor, General Brander.

General Brander's qualifications were different again from those of his predecessors. He was the first Director of Supplies and Transport who was an expert in all aspects of mechanical transport. He joined the Army Service Corps in 1906 and immediately after completing his junior officers' course attended a works course at John Fowlers at Leeds, at that time very much an innovation in the training of an officer of a mounted corps. With this experience he became, at an early age, an instructor at the MT School at Aldershot. During the 1914-18 war he was mainly employed in appointments concerned with the provision and deployment of mechanical transport, including the formation and command of the Base MT Depôt at Calais, after which he held a staff appointment in the QMG 3 branch of the War Office. After the war he returned to Aldershot and, as Chief Instructor at the MT School, played a leading part in research and experimental work. Two projects in which he was deeply involved were the development of the six-wheeled load-carrying vehicle and the giant pneumatic tyre for military use. After service for two years in workshops in India, at Bombay and Chaklala, he returned home to become Chief Instructor of Tracked Vehicles at Woolwich. After a short spell as OC RASC Portsmouth he was appointed Commandant of the RASC Training Centre in 1933 and, three years later, succeeded Brigadier W. N. White as Inspector, in the rank of major general. Throughout his career he had been a single-minded pioneer of mechanical transport as the chief instrument of the Army's mobility.

General Brander's chief task was to maintain the momentum of preparation for the approaching conflict. 1937 was a year of steady expansion. Three additional transport companies were authorised for Egypt – 65, 66 and 67 Companies. The first two were on an establishment of three officers, 64 other ranks and 24 vehicles. 67 Company was designed for troop carrying and had five officers, 220 other ranks and 85 vehicles. At home fourteen new MT Companies were authorised. 1, 2, 3 and 4 Companies were re-formed and number 55 to 64 Companies were formed, bringing the total of companies up to sixty-seven. 68 Company was not formed until September 1938. There was a great increase in the number of recruits at Aldershot and 164 officers joined the Corps (29 Regular, 54 TA and 81 Supplementary Reserve). A second Supply Reserve Depôt, at Barry in South Wales, was planned, and the double echelon supply system was introduced. Establishments of petrol carrying units were worked out and training in anti-gas measures, anti-tank and anti-aircraft procedures, and in supply dropping techniques were everywhere included in exercises.

It was fortunate for the Royal Army Service Corps that, shortly after General Brander's appointment, the Rt Hon L. Hore-Belisha was appointed Secretary of State for War in July 1937. Apart from his service in the Royal Army Service Corps in the Great War he was Honorary Colonel of the 43rd (Wessex) Divisional RASC (TA). He always showed great interest in the Corps and accepted an invitation to be the guest of honour at the Corps' Golden Jubilee Corps Week in 1938.

In the Army Estimates for that year the role of the Army was once more defined in the following terms: (i) home defence, the menace of air attack being a primary consideration, internal security being the second duty in this connection; (ii) the discharge of British commitments overseas, including defended ports on the trade routes; (iii) the use of the 'strategic reserve' after the above claims have been met, whether for the reinforcement of internal security or defence of territories for which we are responsible overseas, or co-operation in the defence of the territories of any allies we may have in the case of war.

To meet these commitments there would be two types of division — motorised and mechanised armoured. The Secretary of State defined the aim of the War Office in the words: 'It will be our ideal in the War Office so to conduct our business that in all important matters there will be speed in decision, flexibility in action, and in administration not only thoroughness but that humanity which can help to give to the soldier in peace-time that security which we expect from him in the blunt eventuality of war.'

The air defence of Great Britain was given the highest priority. It was quickly realised that anti-aircraft formations must have their own supporting RASC units. The 1st AA Div RASC (TA) was formed from 56th Div RASC (TA) when the two London TA Divisions amalgamated in 1936. It consisted of four separate Group Companies, which served their own dispersed AA formations. The 2nd AA Div RASC (TA), consisting of five Group Companies, was formed on 1 April 1938 from existing TA RASC units. Partial mobilisation of Anti-Aircraft Command took place during the Munich crisis, which provided valuable experience in the maintenance of AA formations. By 1939 the number of divisions in AA Command had risen to five and an Assistant Director of Supplies and Transport was appointed to Headquarters AA Command, to co-ordinate the work of RASC support units.

In November 1937 the Mobile Division was formed from the 1st and 2nd Cavalry Brigade (equipped with light tanks) and the Tank Brigade. 10, 27 and 36 Companies were at first allotted to the division and Major P. S. Whitcombe, who had just completed a tour as GSO2 under the Director of Military Training in the War Office, was appointed CRASC Mobile Division in August 1938, with Captain J. A. Mullington as his adjutant. A year later the Divisional RASC was increased by the addition of 4 and 62 Companies. Later still, 11 and 55 Companies replaced 4 and 36 Companies. In Egypt the Mobile Divison (Egypt) was formed and in the Territorial Army the 2nd Cavalry Div RASC (TA) became the Mobile Div RASC (TA).

Two events in 1938 helped to raise the morale of the Corps still higher. On 12 April HM The Queen visited the RASC Training Centre and on 1 August the new 'Belisha' rules for promotion were introduced, based on automatic promotion after a fixed number of years, up to the rank of major. These new rules affected a large number of elderly subalterns and captains.

For some years the Royal Army Service Corps had depended on the output of the Army Technical School at Chepstow for a substantial number of boy artificers. Royal Army Service Corps officers and senior NCOs had always formed part of the instructional and administrative staff of the school. Artificers from this source were of a high standard and most of them earned early promotion to NCO and warrant officer rank. In order to increase the intake of skilled artificers, it was now decided to open a new Royal Army Service Corps Army Technical School for boys, in Jersey under the command of Lieutenant Colonel W. N. Craig-McFeely. This had already become a most successful enterprise when it was forced to close early in the war by the fall of France.

A great deal of what was achieved between 1935 and the beginning of 1939 owed much to the support of the Quartermaster General, General Sir Reginald May KCB KBE CMG DSO. The formation of twenty additional MT companies, the creation of five anti-aircraft divisional RASC (TA) and a greatly improved peace establishment for RASC Territorial Army units received his active support. He gave his encouragement to new measures for the development of a sound system of petrol supply. The appointment, in 1936, of Mr William Fraser (later Sir William Fraser CBE), Deputy-Chairman of the Anglo-Iranian Oil Company, as Honorary Petrol Adviser to the War Office, was an important move with far-reaching effects. The expansion of the staff of the Chief Inspector of Subsidised Transport; improvements in messing and catering and the modernisation of cookhouses and cooking equipment; the appointment of Sir Isidore Salmon as Honorary Catering Adviser to the

Army; improvements to barracks, and in sea transport, with the construction of the *Dilwara*, *Dunera*, *Ettrick* and *Devonshire*, were some of the advances made through his drive and inspiration. He regularly attended the dinners of the Corps Club and became an honorary member.

Throughout 1939, preparations for war accelerated. In March the length of recruits' military training was cut from twelve to nine weeks, in order to increase output, and 35 and 64 Companies, at Aldershot, became driver training companies while at Bulford 47, 48 and 56 Companies were given a similar role. In May the number of officers was increased by the offer of permanent Regular commissions to officers of the RASC Supplementary Reserve and Territorial Army, and by the publication of conditions for temporary Regular commissions. The Territorial Army was issued with its full establishment of vehicles.

On 9 May the Military Training Act 1939 was passed, compelling every male subject between 20 and 21 to undergo six months' training and then to spend three and a half years in the auxiliary force. Those called up under the Act were termed the Militia. This Act was superseded by the National Service (Armed Forces) Act, 3 September 1939 and the National Service Act of 1947.

At Corps Week (15 to 21 May) the Representative Colonel Commandant (General Macleod) announced, in his address on 'W' Square, that the peacetime establishment of the Regular RASC was to be doubled, and that of the Territorial Army more than doubled.

In June, Reservists were called up and compulsory military service came into effect. A Royal Army Service Corps Militia Training Centre (administered by the RASC Training Centre) was established in hutted accommodation in the Ewshott area. The first intake of 660 men arrived on the 1 July. They received two months initial military and regimental training, followed by four months with a working RASC unit.

There was no doubt in anybody's mind that there was not much more time. On 2 September the Feltham correspondent of the Royal Army Service Corps *Journal* wrote:

'And so we carry on with very little to tell you. Crisis succeeds crisis until crisis becomes the new normality, and we become so used to it that one begins to think that an assurance of a real and lasting peace may make life just a little dull!'

In 1941, when he had been succeeded as Director of Supplies and Transport by Major General Snowden-Smith, General Brander became MGA Eastern Command until his retirement. He was then selected by Lord Beaverbrook to become Deputy Director of Mechanisation (Wheeled Transport) in the Ministry of Supply, and so became responsible for the supply of vehicles, their stores and repairs for all the armed Services.

During those years leading up to the war, the QMG branches concerned with supply and transport were staffed by a collection of exceptionally well qualified staff officers, who were largely responsible for most of the detailed preparations which were made in equipping the Royal Army Service Corps for war. At the head of QMG 3 was Colonel E. H. Fitzherbert, with H. C. Goodfellow, H. H. Dempsey and A. R. Purches. In QMG 6, R. T. Snowden-Smith, T. W. Richardson, H. M. Gale, H. M. Whitty, W. H. D. Ritchie and D. H. Bond all played their part. Three of these later became Directors of Supplies and Transport, four became Inspectors and one became a lieutenant general. In other branches of the War Office H. N. Gallagher, L. F. Field, C. M. Smith and R. T. Cooke all held important appointments.

Nor must we forget the great contribution made by that most efficient and intensely loyal body of RASC Headquarter clerks. By the time a Royal Army Service Corps clerk had reached the rank of warrant officer, or had been promoted to commissioned rank, he had accumulated an immense amount of knowledge and experience. In many cases a chief clerk, because he was not always subject to constant moves, represented a degree of continuity denied to the staff officer. Senior Army commanders often found that their chief clerk was a more valuable asset than a young staff officer, who had only recently arrived from the forcing house of Camberley. The combination of a highly experienced chief clerk and the imaginative outlook of a competent trained staff officer gave the best service of all to the senior commander.

'Ted' Collins (Major E. T. Collins MBE) enlisted as a clerk in 1919 and was appointed to QMG 6 on promotion to sergeant in 1926. He remained in QMG 6 for thirty years until his retirement as a major in 1956. He had an unparalleled and encyclopaedic knowledge of everything to do with ration scales, and was a most valued member of the staff of many heads of QMG 6 and, later, of ST6.

The chief clerk (staff sergeant) in a brigade headquarters could be an invaluable asset. In the 1st Cavalry Brigade Headquarters at Aldershot in the early 1930s, Staff Sergeant Talbot RASC was chief clerk. He had made himself so indispensable in everything to do with the brigade headquarters that he was widely known as 'Brigadier' Talbot. This led, on at least one occasion, to confusion and the considerable discomfiture of a young subaltern in the RASC company serving the brigade, who had been detailed by his company commander to make some arrangements for the real Brigade Commander, Brigadier F. W. Bullock Marsham, DSO MC.

Major General C. W. Macleod CB CMG DSO, Director of Supplies & Transport, 1933-1937.

46 & 51 Companies RASC and Supply and Bakery Sections, driving through France in 1935, on the way to the SAAR as part of the British contingent, sent to assist in the Plebiscite.

The Saar Plebiscite XI

The last post-war commitment in Europe in which the Royal Army Service Corps was involved was the Saar Plebiscite at the end of 1934 and the beginning of 1935. Under the Treaty of Versailles the territory of the Saar was placed under the control of the League of Nations from 10 January 1920 for 15 years. In January 1935 the future of the territory would be determined by a plebiscite.

As 1934 drew to a close it was decided to despatch an international force consisting of British, Dutch, Swedish and Italian troops to ensure that the Saar Plebiscite was conducted in a fair and proper manner.

The British contingent of the Saar (Plebiscite) Force — Saarforce — consisted of a composite brigade of two infantry battalions and a cavalry regiment, commanded by Brigadier J. H. T. Priestman DSO MC. The Royal Army Service Corps units involved were: HQ RASC (OC Major E. S. Unwin MC), 46 Company from Shorncliffe (Captain M. M. Simpson MBE), 51 Company from Catterick (Captain N. G. Duckett), and a supply depôt from Aldershot (Major H. M. Whitty and Captain J. K. Gillespie), consisting of a detachment of a supply company and a detachment of a field bakery.

The MT companies were specially equipped with new vehicles, Austin 7 cars, Austin 20 staff cars, Morris Commercial 30-cwt 6 x 4 CD and Thornycroft 3-ton 6 x 4 LF/SC6. 46 and 51 Companies were two of the most recently-formed units of the Royal Army Service Corps, the former having been raised on 1 April 1933 and the latter exactly a year later.

Little time was allowed for preparations for the move, which included the reception of personnel from other companies, the collection of vehicles, and the assembling of all the additional stores required for an indefinite period overseas. The warning order was received at 7 pm on Monday 10 December. The message received by Captain Simpson at Shorncliffe was: '46 Company selected for duty in the Saar, will probably leave at the end of the week.'

51 Company marched out of Catterick early on 13 December and arrived at Feltham in the evening of the following day. The balance of the company's vehicles were issued from the Vehicle Reserve Depôt and on Sunday 16 December the company marched from Feltham to Shorncliffe with a total strength of 108 all ranks, six motor-cycles, 33 motor cars, 24 lorries and 13 trailers. The vehicles were embarked on the 17th and arrived at Calais on the 18th. After two stops in France, at Douai and Mézières, 51 Company, which had been joined by 46 Company at Calais, arrived at Saarbrücken on the 21st — ten days after the receipt of the warning order. After a short rest the Company was moved to the northern area of the Saar territory, allotted to the 1st Battalion The Essex Regiment, and billeted in the mine buildings at Schiffweiler, adapted for their use. Captain Duckett had two subalterns with him, Lieutenant E. C. C. Lewis and Lieutenant J. B. Sandberg.

46 Company was made up to establishment between Wednesday and Saturday by a steady stream of men and vehicles from Bulford, Chatham and Woolwich. The establishment of officers was completed by the arrival of Lieutenants C. V. Ferrey, T. E. Southwell and J. G. C. Low from 22 Company, 21 Company and 49 Company. The workshop and store lorries were unwrapped from their winter jackets and by Saturday the company was ready to move.

For the journey from Calais to Saarbrücken the French Army provided every assistance, including the loan of two liaison officers from the 19e Escadron de Train, Commandant Mouilban and Lieutenant de Montjamont. The civilian population was lavish in its welcome and hospitality. This was the first time that Royal Army Service Corps units had been on the Continent since the evacuation of the Rhineland more than five years earlier. When the convoy of 160 vehicles arrived at Metz the reception was tumultuous. Bands played the British and French national anthems and many of the familiar wartime songs, and the gendarmes were hard put to it to keep the crowds under control. On the door of the men's mess room the bill of fare read:

Friday – Dinner
Potage Vermicelle
Saumon Vinaigrette
Roast Beef
Pommes Rissolés
Confitures – Madeleines
Café
Bordeaux Blancs
Champagne

Saturday — Breakfast
Thé ou Café au lait
Jambon — oeufs — confitures
Petits-pains

After the French-Saar border was crossed the contrast was quite shattering. The convoy's progress was watched in stony-faced silence. In Saarbrücken there were still the same crowds, but this only served to emphasise the silence of their 'welcome'. Thousands of red banners with a black swastika on a white ground hung from walls and windows.

The supply detachments, which had been assembled at Aldershot on 13 and 14 December, crossed the Channel on the 18th and, after entraining at Calais, arrived at Saarbrücken on 19 December and were conducted to their billets in a huge railway engine workshop (Werkstätte) which, in addition to accommodating 46 Company, was to serve as supply depôt, sleeping and messing quarters. After strenuous efforts, which demanded a considerable amount of improvisation, the supply section and bakery were set up in time for issues on 21 December.

Voting day for the plebiscite had been fixed for Sunday 13 January 1935 — referred to by everyone in the Saar as 'Der Tag'. For the Saarlanders there was no other topic of conversation. They took a great deal of interest in the role of the peace-keeping force. Their attitude was mainly one of friendly tolerance.

The responsibility of the International Force was two fold. First and foremost they had to keep the peace and to prevent rioting in the streets. There had been a considerable amount of 'Nazi' propaganda and the 'Deutsche Front' organised large scale demonstrations and processions (estimated as up to 100,000 strong). On the whole these were non-violent and adequately controlled by the international police; sometimes they made the movement of vehicles difficult. The second role of the force was to ensure that voting was carried out in an orderly manner and to collect the urns after polling had been completed.

During the fortnight before voting took place, one of the main activities of the Royal Army Service Corps companies was to reconnoitre routes and to learn the location of the polling stations. It was also necessary to make close contact with the infantry battalions with which they would be working and to perfect embussing procedures for the efficient performance of 'riot duty'.

By the evening of Friday 11 January there was a feeling of confidence that all necessary preparations had been made. By the following morning the situation had been completely altered by a heavy fall of snow, which continued most of Saturday morning. Special arrangements were hurriedly made (Captain Simpson even went so far as to telephone Feltham about the availability of chains), but in the event all went reasonably well, although some of the convoys collecting urns had a little difficulty in getting through to the special Dutch train arranged for their delivery to the counting centre. A rum issue at the end of the day was welcome.

The result of the plebiscite was broadcast on Tuesday 15 January and showed an overwhelming vote for return of the territory to Germany. Of the 525,758 votes cast (98% of the electorate) 477,119 (or 90.3%) were in favour of Germany, 46,513 supported the status quo and only 2,126 were for union with France. Nazi flags and swastikas appeared everywhere; shops were closed and, amid general rejoicing, the day was taken as a holiday. But this was not the end for the peace-keeping force, which was to remain for another six weeks.

The main task accomplished, RASC companies concentrated once more on individual training, so rudely interrupted in December, and on what recreational training could be arranged with other units, the police and civilian teams.

When word came in the middle of February that the force was to be withdrawn by the end of the month, there were few who did not regard it as welcome news. The return through France was as well-organised by the French Army as the outward journey. By 3 March both companies and the supply detachments were home again, after a three month continental interlude which had provided a valuable (and often exciting) experience.

The following extract is from a letter written by the brigadier commanding the British contingent to the Director of Supplies and Transport:

'I hope you do not mind my writing to you about the detachment that was with me for the last two and a half months; you will have heard quite a lot about them; but whatever you have been told cannot be quite all. I doubt if there was ever a closer or happier touch between the RASC and the infantry; in fact they were just one, and this went down to ceremonial parades. We never had any breakdowns, and on the big occasion when the roads were grim, the young drivers rose to the occasion like veterans. Conduct admirable – everyone cheerful. I have had a very real affection for the corps all my service, since the days they were so kind to me at Aldershot, and it was very delightful once more to be alongside the same old spirit I have always known.'

The White House, the RASC Officers' Mess, Abbassia, Egypt. (PGT)

British Troops in Egypt XII

Up to the outbreak of the 1914-1918 war Turkey maintained her nominal suzerainty over Egypt, in spite of the virtual independence established by Mehemit Ali in the 19th Century and the presence of a British occupying force. On 20 December 1914 the pro-Turkish Khedive was deposed and a British Protectorate declared. Although Egypt was the main base for Allied operations in the eastern Mediterranean, it remained nominally neutral. Wartime conditions and the presence of large numbers of British and Dominion troops tended to antagonise growing nationalist elements, and early in 1919 there was serious anti–British rioting. Zaghlul, leader of the emerging nationalist party, did his best to foster disturbances in Cairo and provincial towns like Tanta, inciting his unruly followers to mob violence, disrupting communications and damaging the economy. Most of the occupying troops were in camps around Kantara, well away from the main centres of population, and not directly involved.

Zaghlul was deported to Malta, and in March 1919 Allenby was appointed High Commissioner to restore order. A month later President Wilson recognised the British Protectorate of Egypt and soon after Allenby, rather surprisingly, decided to release Zaghlul. The outrages continued, but without involving the occupying troops, whose thoughts were mainly fixed on their impending return home. The Administration was fortunate in having Sir Russell Pasha as Commandant of the Cairo Police. He fully realised that the use of troops would only exacerbate the situation and goad the agitators, who were mostly students from El Azhar, to even more violent actions. Succeeding High Commissioners heeded his pleas for non-intervention and left him a free hand. He understood the Egyptian mentality and, by his firm but unobtrusive methods succeeded in avoiding heavy bloodshed. At times the situation looked extremely ugly, and Egypt's economy suffered severely from wanton destruction. All through 1921 the disturbances continued and Zaghlul's irresponsible behaviour led once more to his deportation — this time to the Seychelles.

In the hope of creating a more favourable atmosphere for negotiation, Allenby declared the Independence of Egypt on 20 February 1922, and a month later Sultan Ahmed Fuad was proclaimed king — on 16 March.

The declaration of independence was limited by four reservations, which it was intended should be resolved by future agreement. The presence of the British in Egypt was to protect what the Egyptian Government and the other interested powers recognised as their legitimate interests: (i) the security of the communications of the British Empire which crossed Egyptian territory and in which the British Government held a large financial interest; (ii) the defence of Egypt against foreign aggression or interference; (iii) the safeguarding of foreign interests in Egypt and the protection of minorities; (iv) the administration of the condominium of the Sudan, under the Anglo-Egyptian agreement of 19 January 1899.

In 1923 martial law in Egypt came to an end and Zaghlul once more returned from exile. As might have been expected, nationalist agitation continued, fanned by the Wafd party, until democratic elections were promised. The country was hardly ready for them. There was little political consciousness among the population, except in the ranks of the nationalists, and they – the Wafd Party – predictably secured an enormous majority.

King Fuad had no choice. In January 1924 he appointed Zaghlul Prime Minister. His tenure of office coincided almost exactly with that of the first Labour Administration under Ramsay MacDonald, and ended as abruptly.

Zaghlul, in spite of his position of authority, continued to encourage violence among his supporters and to foster discontent in the Sudan. These disturbances reached their climax with the assassination in Cairo of the Sirdar, General Sir Lee Stack, who was Governor General of the Sudan and Commander in Chief of the Egyptian Army, on 19 November 1924. Allenby's reaction was swift and resolute. He demanded an unequivocal apology from Zaghlul, an indemnity of £500,000, and the immediate submission of the Egyptian Government to British policy in the Sudan. Zaghlul accepted the terms of Allenby's demands without protest and immediately resigned. He died three years later at the age of 74.

Colonel H. G. Burrard CB CSI DSO ADC, who was ADS&T, British Troops in Egypt, was one of the pall bearers at the Sirdar's funeral.

One of the immediate effects of the assassination of the Sirdar was the creation of the Sudan Defence Force. Up to 1924 the Egyptian Army, which was officered by British officers and NCOs, served in the Sudan as well as in Egypt. One of the conditions imposed on the Egyptian Government after the murder of General Stack was that all units of the Egyptian Army would withdraw from the Sudan.

A treaty was negotiated in 1927 but, owing to the opposition of the Wafdist Party, it was never ratified. Another nine years were to pass before the Anglo-Egyptian Treaty of 1936 was finally agreed on 26 August, four months after the accession of King Farouk, the 16 year old son of Fuad. The Italian invasion of Ethiopia in 1935 had obliged the Egyptians to face the brutal facts of international politics and a 'United Front' of Egyptian politicians agreed to reopen negotiations.

This Treaty recognised the sovereignty and independence of Egypt and provided for an alliance for twenty years with Great Britain. Great Britain undertook to withdraw troops from the Delta to cantonments in the Canal Zone, which they were to occupy and defend until the Egyptian Army had been modernised and re-equipped, with British technical advice. The number of British personnel in the canal zone would be limited to 10,000 (land forces) and 400 pilots (air forces), together with the necessary ancillary personnel for administrative and technical duties. There were other provisions for the allotment of training areas and the improvement of communications. The Condominium of the Sudan was to be maintained. Britain would support Egypt's application to join the League of Nations. (She became a member in the following year.) The 'Capitulations' were abolished and the Egyptian Government would be exclusively responsible for the lives and property of all foreigners in Egypt. British personnel were withdrawn from the Egyptian Army and a Military Mission was established without delay; modernising the Egyptian Army went ahead. Little progress had been made in building alternative accommodation in the Canal Zone beyond the drawing board stage, when war broke out in 1939.

When the British Army withdrew from occupation of the Rhineland and the Royal Army Service Corps handed over responsibility to the Indian Army Service Corps, Egypt became the largest overseas station for the Corps. The British troops in Egypt, with HQ BTE in Cairo, were mostly garrisoned in barracks in Cairo, and in the neighbouring suburbs of Abbassia and Heliopolis. The peacetime force consisted of a light brigade of three mechanised cavalry regiments, one equipped with light tanks, one with armoured cars and one with pick-up trucks, a Royal Tank Corps regiment, two Royal Artillery regiments and five Infantry battalions, forming the Cairo Brigade and the Canal Brigade, with Royal Engineer, Royal Corps of Signals, Royal Army Service Corps, Royal Army Medical Corps and Royal Army Ordnance Corps units.

In addition to Cairo, the Egypt Command included the garrisons of Alexandria, Moascar (the military cantonment outside Ismailia on the Suez Canal), Port Said, Suez, Khartoum and Cyprus, at all of which there were Royal Army Service Corps detachments. At Port Said and Alexandria the embarkation staffs were found by the Royal Army Service Corps, as they were at Southampton and in the Far East stations. The command was also responsible for providing troops for the maintenance of law and order in Palestine, until it became a separate command during the Emergency in 1936. To meet all these commitments the RASC units consisted in 1927 of 25 (HT) Company, 30 and 31 (MT) Companies, and 'D' Supply Company, which had previously been 23 (LofC) Supply Company. All these units were stationed in Abbassia under the command of the OC RASC Cairo District, in well built barracks. The officers' mess was in the 'White House', which had once been Kitchener's residence. The oak panelled walls of the ante-room were later decorated with a collection of life-like cartoons of officers who passed through the mess, skilfully drawn by Lieutenant J. F. 'Tishy' Vaux. The two Sudanese mess servants, Abdul and Rahoun, were well known to more than one generation of Corps officers.

On 1 October 1928, 25 Company was converted to MT, and in March 1929 a new company was formed for service in Egypt — 39 Company, which was organised as a mobile repair unit.

The peace establishment of three service MT companies then amounted to less than 100 30-cwt lorries, which barely met the day-to-day requirements of the Cairo garrison and did not allow for decentralisation to the Command's out stations. In Moascar, where the headquarters and two battalions of the Canal Brigade were stationed, and Alexandria, where the third battalion was located, the OC RASC had to depend on unit transport or, if this was insufficient, hired vehicles. The requirement for transport in the smaller places was minimal. For example, in Port Said the RASC detachment consisted of one captain, who was OC troops and OC RASC details, and two subalterns, one of whom was the embarkation staff officer and the other the transport officer, together with a small establishment of clerks. The garrison in Cyprus consisted of one infantry company, detached from the Canal Brigade, which was located in Polymedia Camp, four miles north of Limassol, with a summer station in Troodos — a much sought after assignment. The RASC detachment was one of the smallest in the world, consisting of only two military personnel, a warrant officer and a clerk, and a civilian expense store accountant, based at Limassol, until the arrival at the end of 1934 of four Morris 30-cwt 6-wheeled lorries increased the establishment by 300 per cent. The 'rebellion' in 1931 had no doubt shown the inadvisability of relying entirely on hired transport in an emergency. In the same year the garrison was moved from Limassol to Nicosia, the seat of government and a more central position.

The small RASC detachment in Khartoum was administered by Headquarters British Troops in Egypt but came under the command of the GOC British Troops in the Sudan (a two battalion command), who was also Commandant of the Sudan Defence Force – a Foreign Office appointment.

When transport units could be spared from garrison administrative commitments, Egypt offered excellent opportunities for company training. The open desert was easily accessible and consisted of a great variety of surfaces, from soft sand to hard stony gravel, from virtually impassable sand dunes to low tussocky scrub which, although it was no serious obstacle, made driving difficult except at low speeds. Drivers soon learnt to overcome these hazards and with the use of sand-channels and sand-mats, which had been invented by Major (later Brigadier) R. A. Bagnold in 1929, became expert at negotiating all types of going, including wide stretches of soft sand. The absence of conspicuous landmarks and the effects of mirage during the heat of the day made normal map-reading impossible. Navigation depended on dead-reckoning and constant use of the compass. As there were no compensated compasses provided for RASC vehicles, the prismatic compass was virtually useless, as true readings could only be obtained at some distance from a vehicle. Early pioneers in desert navigation, notably Brigadier R. A. Bagnold, had evolved a sun-compass, which overcame these difficulties and was extremely accurate over long distances. In a country where the sun always shone this was a useful piece of equipment. A simple model was developed which could be easily fittted to any vehicle and this made navigation in a featureless landscape comparatively fool-proof.

Lack of features made concealment from the air a difficult problem. Sand-coloured paint made an excellent camouflage, but it could not conceal the hard dark shadow cast by a lorry. This was not serious at mid-day when the sun was overhead but became increasingly noticeable when the sun was low in the sky. What was also likely to betray a vehicle was the plume of sand dust which each lorry trailed behind it as it crossed the desert. Under certain conditions vehicle tracks through soft sand could also be picked up from the air. 208 (Army Co-operation) Squadron RAF was very helpful in bringing home these points to company commanders.

It was soon learnt that the best defence against air attack was wide dispersion and immobility. By turning a lorry away from the sun, the size of the shadow was reduced to a minimum, and by avoiding any sort of regular pattern quite a good degree of concealment could be obtained.

Most men thoroughly enjoyed these company 'outings' in the desert, which tested their driving ability and their resourcefulness. Ten days or so away from the noise and bustle of Cairo in the clean air of the desert was a welcome change and not long enough for men to become bored or disinterested.

Until the 6-wheeled 30-cwt became widely available, a feature of collective training in Egypt was maintenance by camel transport. In 1928 and 1929, brigade training in the desert between Mena and Sakkara involved the use of as many as 1,700 camels, and in 1933 a camel transport unit of 1,000 camels and 500 native camel-drivers maintained a force of 4,500 men and 800 animals for three days with supplies and water.

The carriage of water was always a major problem in desert warfare. In 1931, for the first time, a feature of collective training was the delivery of water by mechanical transport. Four Morris 6-wheelers were fitted with two water tanks each of 150 gallons capacity, and four Thornycroft 6-wheelers carried 25 'fanaties' each. A 'fanatie' held 12½ gallons of water and was the standard camel water container, two of them constituting a camel load. This small detachment could carry 2,450 gallons of water.

There were a number of trials, some of them semi-official, to test the desert worthiness of mechanical transport under difficult conditions. One, in which Captains C. F. Neve and L. A. Tudor took part, went to Bir Hooker in 1931, and in 1933 an expedition from Cairo to Siwa Oasis across the Libyan Desert was organised by the 12th Royal Lancers with armoured cars, including detachments of RSignals, RASC, RAMC and RAOC. A more ambitious trial, sponsored by Dunlop, was undertaken in 1938, with a number of RASC officers from the Inspectorate of Mechanical Transport taking part. Its main object was to test the behaviour of different tyres under strenuous desert conditions.

During the early 1930s the Corps was fortunate in having Brigadier J. C. Browne CB CMG DSO ADC as brigadier in charge of administration at Headquarters British Troops in Egypt.

The move, at short notice, of 25 Company to Palestine in August 1929 revealed the lack of any transport reserve in Egypt to deal with emergencies, but it was not until 1935 that it was decided to re-form 5 Company for permanent service in Egypt. The decision had hardly been taken when the Italian invasion of Ethiopia precipitated an emergency which necessitated the immediate despatch of units from the United Kingdom. In addition to 5 Company, four other companies, 17, 19, 49 and 51, were sent, and nearly thirty other officers were posted for temporary duty.

This emergency and the deteriorating political situation in Europe provided a stimulus to further mechanisation. The Cavalry Brigade was reorganised as the nucleus of a mechanised mobile division. 5 Company took on the temporary role of a driving school to assist in the training of drivers of other arms, and in 1937 authority was given for the formation of three additional RASC Transport Companies in Egypt, to meet the demands of training and the requirements of the garrison.

65 and 66 Companies were formed before the end of the year, on the same establishment as 5, 25 and 30 Companies, with three officers, 64 other ranks and 24 vehicles. 67 Company was to be a troop carrying company with five officers, 220 other ranks and 85 vehicles, taking over the role and most of the personnel and equipment of 19 Company, on its return to England.

The RASC units in Egypt at the beginning of 1938 amounted to: 'D' Supply Personnel Company (the new title of a Supply Company since May 1937) and 5, 25, 30, 31, 39 (MRU), 65, 66 and 67 MT Companies. This increase made it possible to send 25 Company, under the command of Captain A. F. J. Kingsmill, to Moascar for permanent duty in the summer of 1938, and 67 Company later in the year on a short spell of temporary duty in Palestine.

The Munich crisis of September 1938 was the signal for the Mobile Desert Force to move out to Mersa Matruh. Brigadier H. E. Russell was in command of the force. In addition to the Cairo Cavalry Brigade (consisting of the 7th, 8th and 11th Hussars) the force included the 3rd Regiment Royal Horse Artillery, the 1st Battalion Royal Tank Regiment, the 2nd Field Squadron Royal Engineers, 5 Company Royal Army Service Corps and the 2/3 Cavalry Field Ambulance. 5 Company was commanded by Captain S. M. Griffiths.

When the crisis ended on Neville Chamberlain's return from Munich, the Mobile Force returned to Cairo, where Major General P. C. S. Hobart had just arrived to form and command the Mobile Division (Egypt). With the addition of the 6th Battalion Royal Tank Regiment and the 1st Battalion the King's Royal Rifle Corps, a motorised battalion, the force was developed in a remarkably short time into the Mobile Division, which subsequently became the 7th Armoured Division.

In the early part of 1939, four RASC companies were allotted to the Mobile Division — 5, 25, 65 and 66 Companies. These companies between them could muster 96 2-ton 'Leader' Morris lorries. This was the number which went with the division to Mersa Matruh in August 1939. Major W. J. F. Eassie arrived in Egypt in the summer of 1939 and replaced Lieutenant Colonel J. T. Field, who had been acting as CRASC of the Mobile Division.

Later on, after the outbreak of war, 25 and 66 Companies were replaced in the 7th Armoured Division by 58 Company, which had come from Palestine in July 1939, and 550 Company, the non-divisional company of the 48th (South Midland) Divisional RASC (TA) which, after mobilisation, had been sent out from England with the 1st Cavalry Division to Palestine.

Major C. M. Smith MC, had arrived from England slightly ahead of General Hobart, to be DAA and QMG Abbassia from the 1 September 1938. Just before Christmas he became DAA and QMG, Mobile Division, British Troops in Egypt.

Major General Verney DSO MVO, in *The Desert Rats* wrote: 'As his right-hand man General Hobart had Major C. M. Smith, RASC, and together they evolved the administrative basis on which the division was to fight successfully in the coming war'.

The main feature of desert administration which was evolved was the grouping of refilling points into a replenishment area, where second-line transport transferred its loads to 'B' echelon vehicles, at a predetermined place and hour in the shortest possible time. Frequent exercises were held during 1939 to practise and perfect this system, which depended on good wireless communications and carefully controlled dispersion. The system fully proved its worth in subsequent operations in the desert.

No one was deceived by the claim that peace had been won at Munich. At best it provided a breathing space and an opportunity for further preparation.

At the end of 1938 an up and coming young major in the 5th Inniskilling Dragoon Guards, Major C. Keightly, gave an eloquent account to all the officers of the Abbassia garrison of his experiences during a recent attachment to the German Army. Nobody was in any doubt, after this electrifying talk, that it would not be long before we would be at war.

During the remaining year a great deal was done to improve the resources of Egypt as a base for operations in the Middle East. Stocks of food stuffs and of petrol in 4-gallon containers were built up in dispersed depôts to avoid the danger of air attack, and preparations were made with the oil companies at Suez to meet the Army's requirements. Supply depôts were established at El Dabaa and Mersa Matruh, which were maintained by direct rail from Alexandria.

When it was clear there was no immediate threat from the Italians, most of the Mobile Division returned to Cairo, leaving only reconnaissance forces in the Western Desert to watch the frontier. Thus a further respite was granted, a time of great value in the continuation of training and the advancement of administrative preparations for the struggle ahead.

The Italian invasion of Ethiopia in 1935, which was followed by sanctions against Italy, posed a threat to the whole British position in the Sudan and Egypt. Economically, Ethiopia has a special importance to these countries, because the greater part of the flood waters of the Nile have their source in Lake Tana and the Blue Nile.

It was therefore deemed imperative to reinforce the British troops in Egypt, as a precaution against any hostile acts by the Italians.

5 Company, which had been disbanded at Aldershot on 30 August 1928, shortly after its return from the Rhine, was re-formed on 16 October 1935 at Woolwich – its old pre-1914 home, when it was well known as the 'Roans and Greys' – and embarked on 1 November 1935 for permanent peacetime service in Egypt, stationed at Abbassia. Captain F. H. E. Whittaker commanded the company and his other officers were Lieutenant C. V. Ferry and Lieutenant P. B. Huxham. 'Tiny' Huxham presented problems for the embarkation staff at Southampton, as there were no berths in the troopship which could accommodate an officer who was 7 ft 2 in tall.

At the same time 17 Company from Woolwich (Major C. J. Williams) and 19 Company from Catterick (Major H. M. Hinde) sailed for Egypt in HT *Somersetshire* on temporary duty. 19 Company was put on a special establishment of six officers and 281 other ranks, and 193 vehicles which had been shipped in advance.

A fortnight later they were followed by 51 Company (Captain A. L. Taylor) and, before the end of the year, by HQ 5th Divisional RASC (Lieutenant Colonel A. B. P. Pereira DSO) and a field ambulance.

They were joined in February 1936 by 49 Company from Portsmouth. This company was equipped with 3-ton Albions and Leylands and manned by reservist drivers, called up for the emergency. This was the second time that reservists had been called up since the war, the first time being the Shanghai emergency.

In addition to those on the establishments of companies, between twenty and thirty other officers were posted to Egypt for temporary duty.

17 Company was stationed in temporary accommodation at Abbassia, in Cambrai Camp, together with 2/3 Cavalry Field Ambulance and the transport wing of a motor ambulance convoy. All the other units, including 5 Company, went to Mersa Matruh as soon as they had been issued with their vehicles, where they remained until the middle of 1936.

The completion of the conquest of Ethiopia in May 1936 removed the immediate threat in the Western Desert and the need to keep troops there. Far away in Somaliland the strain on Lieutenant Colonel C. V. Bennett OBE RASC, the commander of the Somaliland Camel Corps, and whose territory was uncomfortably close to the Italian advance, was also considerably eased.

While they were at Mersa Matruh the three companies had plenty of opportunity for training in the desert: driving over a variety of surfaces between Mersa Matruh and Sidi Barrani and up to the frontier at Sollum; map reading by dead reckoning; and the use of sand mats and sand channels in driving through soft sand.

Some of those who spent several months in and around Mersa Matruh during the Abyssinian emergency would have been surprised to read what H. V. Morton wrote in the *Daily Herald*:

'Having arrived at Mersa Matruh in the dark, as I described in my last article, I was unprepared for the sight which met my eyes in the morning. No one had told me that this remote meeting place of desert and sea is one of the most beautiful sights in Egypt.'

Apart from swimming there were not many recreational facilities at Mersa Matruh. 2nd Field Company Royal Engineers had, it is true, built a squash rackets court in the grounds of the Governor's residence in 1936, as a plaque above the door bore witness. It was not much use to the Egyptian Governor, who knew nothing about the game, but who insisted on watching two energetic young subalterns playing from inside the court, a situation which caused as much embarrassment to the players as to the Governor, who was frequently struck. A few years later an enquiry came to the OC 2nd Field Company from the Exchequer and Audit Department, or some such body, asking under what authority a squash rackets court had been built in such an out of the way place. The company commander felt it safe to reply that there was no squash rackets court at Mersa Matruh. The reference in the accounts merely referred to a rendezvous used for local employees, which was an old derelict shell of a building which *looked* like a squash rackets court.

Although the Italian emergency was over, another trouble spot was boiling up. Since the troubles in 1927 Palestine had been comparatively quiet, but under the surface the irreconcilable elements of Jewish and Arab nationalism and racial and religious antagonism had been steadily growing. So, instead of going home, HQ 5th Divisional RASC and 17, 19 and 49 Companies found themselves on the way to another emergency – this time in Palestine – while 51 Company remained in Egypt until January 1937, when it returned home in HT *Neuralia*.

ABOVE: Lt Colonel W. R. V. Warren OBE MC RASC in the White House reflecting on the problems of Desert warfare? (JFV) BELOW: There is a fly in my soup! Major C. H. Moore MC berates the mess staff of the White House; L to R: The PMC, Capt E. H. Carlton RASC; Rahoun; Abdul; Cpl Porter, the Mess Corporal. (JFV)

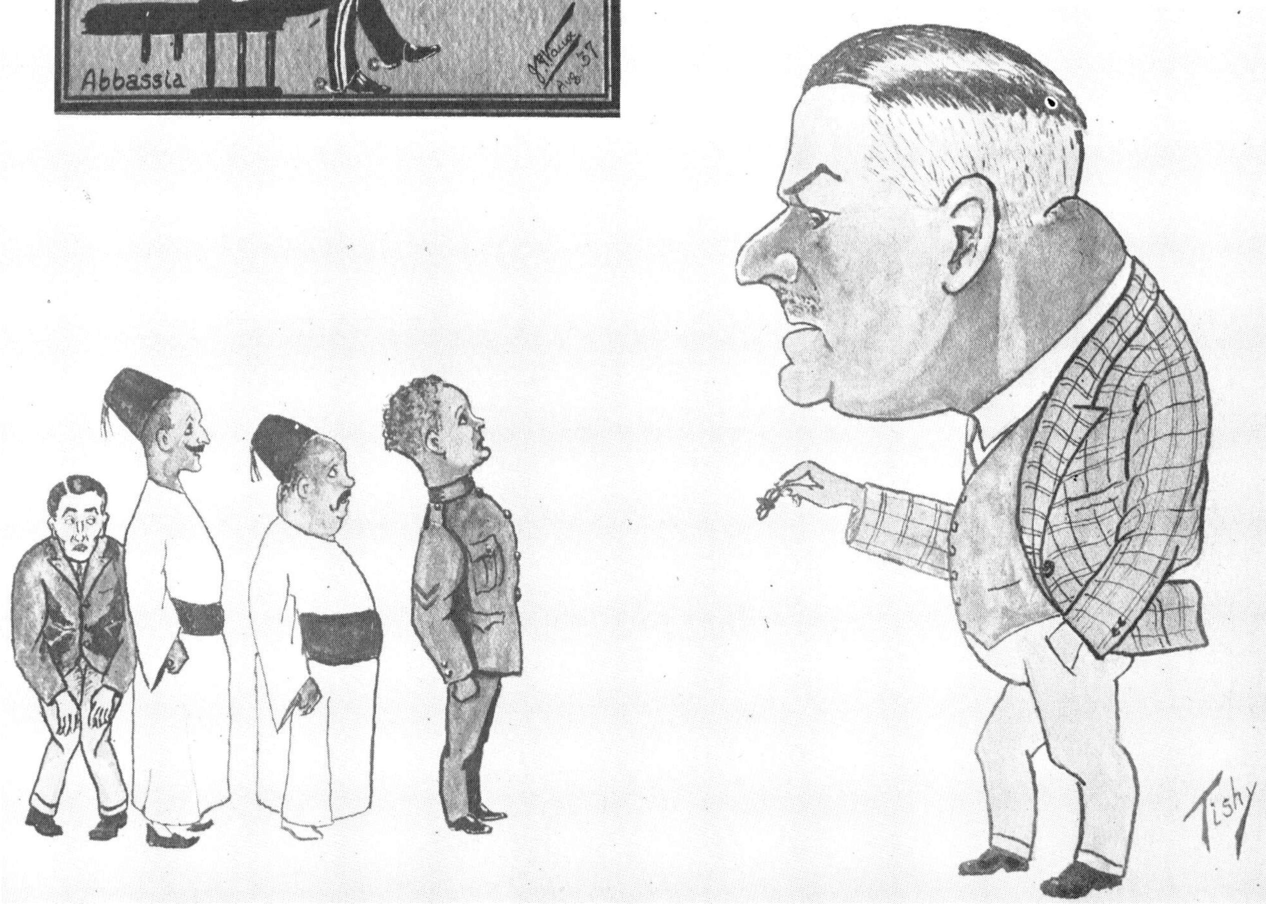

ABOVE: The Officers, Warrant Officers, Staff Sergeants and Sergeants, RASC Alexandria, Egypt 1924; top: SSgt Moss; Sgt Barker; Sgt Lord; Sgt Buckland; LSgt Miles; Sgt Brownrigg; middle: Sgt Chisnal; Sgt Mayes; SQMS Ormes DCM; Sgt Elphick; SQMS Rookes; Sgt Mendham; SSgt Eagle; CQMS Hollamway; front: MSM Fearn; Capt QM Sidwell; Capt Richardson OBE; Lt Col McGwire CBE; Capt Hime Lt Acheson; CSM Holliway. BELOW: Sudan Detachment RASC 1932; seated: LSgt C. Abbott; SQMS C. W. Fletcher; Lt A. W. Flood; Capt C. Boon; Colonel W. P. Robinson DSO OBE; Lt J. C. Bates; Lt W. H. Western; SSgt T. E. Woolgar; SSgt H. E. Burden.

Move of 5 Company RASC to Palestine in September 1936. Defective steering caused this Karrier to run off the road and fall 6ft into a wadi bed on the Nablus-Tulkarm road.

The Palestine Dilemma XIII

The Jewish-Arab problem can be said to have originated in the Balfour Declaration of 2 November 1917, when the British Secretary of State for Foreign Affairs announced:
'His Majesty's Government views with favour the establishment in Palestine of a national home for the Jewish people . . . it being clearly understood that nothing shall be done which may prejudice the civil and religious rights of existing non-Jewish communities in Palestine, or the rights and political status enjoyed by Jews in any other country.'

Two other declarations of policy had been made during the war, when the disintegration of the Ottoman Empire appeared to be imminent, but before the actual liberation of the Arabs from Turkish suzerainty had taken place.

The MacMahon pledge, worked out in Cairo, when Sir Henry MacMahon was High Commissioner, gave promises of Arab independence in the liberated territories, with reservations over the Mediterranean coastal strip, where the populations were not entirely Arab. The Sykes-Picot agreement, negotiated by the Arab Bureau — an intelligence organisation which had been set up in Cairo — with the French authorities, broadly defined the future boundaries of British and French spheres of influence in the Middle East.

Although there were inconsistencies in these three statements of policy, they were overcome when the League of Nations mandates for the Middle East were eventually issued. But there remained considerable scope for misinterpretation – often deliberate – as the conflict of interests in the mandated territories became more acute.

There had been Jewish settlements in Palestine before the Balfour Declaration. The concept of Zionism had been formulated in Russia in about 1860. It was formed into an international organisation by Dr Herzl of Vienna in 1897. With financial support from the Rothschilds, seven farm colonies of Zionist refugees had been settled in Palestine before the end of the century. By 1914 there were about 80,000 Jews in Palestine, nearly all of whom were concentrated in rural enclaves, where they led a communal, pastoral life, which presented no foreseeable menace to the surrounding Arabs.

The Balfour Declaration, the outcome of suggestions and pressures from two prominent Jews, Sir Herbert Samuel, the English Liberal, and Dr Chaim Weizmann, the industrial chemist, clearly envisaged a Jewish National Home, not a Jewish state in Palestine.

The position in 1918 was not clear. Having liberated the Arab territories from Turkey, Great Britain had the responsibility of maintaining law and order and of protecting civil and religious liberties in the conquered areas. For this purpose a Military Government was set up by Allenby. In April 1920 the Allied Supreme Council, meeting at San Remo, ratified the policy of the Balfour Declaration and decided that Great Britain should be entrusted with the Palestine mandate, which was not finally issued by the League of Nations until 1923.

Meanwhile, in June 1920, a civil administration for Palestine, with Sir Herbert Samuel as High Commissioner, was set up in place of the Military Government. The form of government was that of a Crown Colony, with a nominated council and a civil service, in which the higher appointments were filled by British officials.

The Palestine Police, which was an exceptionally efficient and well organised force, was trained on military lines and consisted of some 700 British and 1,600 locally enlisted personnel. Many of the officers and NCOs came from the Royal Irish Constabulary, which was disbanded on the creation of the Irish Free State in 1922. It also included a good proportion of ex-Army personnel who joined the force after demobilisation in 1919. Some of those who served in Palestine during the war (like the Nairn brothers and the Spinneys) saw that there would be opportunities for initiative under a new administration.

For the first five years all went well. Civil administration was greatly improved, in law and order, in health, in education and in public works. Land settlement schemes were developed. Hebrew was revived as a living language.

A steady stream of immigration began to flow, both openly with official sanction, and clandestinely. Some 25,000 agricultural workers were settled in about 100 village communities (at a cost of £6,000,000 to Zionist organisations) and 30,000 immigrants developed the village of Tel Aviv into a Jewish city, with municipal self-government.

But the Arabs were not happy. Theirs was a parallel case with that of the Malays. Both were happy go lucky races, who were prepared to sell their land for the prospect of short term gains to the immigrant Jews or Chinese. But when they saw how the new owners reaped the rewards of their unremitting industry, and looked with envy at the flourishing orange groves or the green paddy fields, they began to grumble and to complain that they had been robbed of their birthright. As the tide of immigration continued to rise the Arabs began to fear that, in time, they would be outnumbered and they might find themselves as a struggling minority in their own country.

Apart from the task of building up the economic life of the country, the High Commissioner's main concern was to reconcile the national aspirations of the Arabs with the effects of Jewish immigration, the main ones being the increase in Semitic influence and the occupation of large tracts of land. The problem became increasingly difficult as existing religious animosities were inflamed and inter-racial jealousies came to a head.

Sir Herbert Samuel had tried to appease Arab sentiment by pointing out that the influx of Jewish wealth brought great benefits to all members of the community. He convened a general Moslem Council as a counter weight to the Jewish Agency, the channel of communication between the immigrant community and the Government, but all to no avail. The Arabs elected as their President Haj Amin el–Hussein, the Mufti of Jerusalem, who was completely unco-operative, denied the validity of the mandate and bitterly opposed Jewish immigration.

One of the first signs of trouble was the general strike called by the Arab community in 1925, when Lord Balfour inaugurated the Hebrew University in Jerusalem. Four years later, in 1929, anti-Jewish activity erupted more violently. There were ugly incidents at the Wailing Wall in Jerusalem, and riots in other parts of the country indicated that the task was beyond the resources of the Palestine Police, and made it necessary to send military reinforcements from Egypt and Malta.

On 24 August the 1st Battalion The South Wales Borderers was despatched from Egypt to Jerusalem with a small supply detachment of the Royal Army Service Corps under Lieutenant L. O. Joliffe. Two days later the 2nd Battalion The South Staffordshire Regiment sailed in HMS *Courageous* from Malta. Two other battalions were placed under orders to move at short notice. On the same day 25 (MT) Company RASC entrained at Abbassia (Cairo) for Palestine under the command of Captain H. M. Cadic. The company was 56 strong and was equipped with twenty Thornycroft and three Morris 6-wheeled lorries, one car, two motor-cycles and ten trailers. They were followed by six Morris 6-wheeled ambulances. The force was maintained by rail from the supply depôts at Abbassia and Moascar to the railhead at Lydda twice weekly.

The rapid despatch of troops to Palestine effectively stifled growing unrest and, after a month, when things had quietened down, 25 Company was able to return to Cairo on 21 September. The supply detachment remained in Palestine, but moved to Sarafand, where the Royal Air Force had its supply depôt and a general hospital.

The situation remained reasonably quiet for the next two years but, as it seemed likely that at least two battalions would remain in Palestine for some time, it was decided to send a Royal Army Service Corps Company from England, to be stationed there. 14 Company was selected and sailed from Southampton in HT *Lancashire* on 10 January 1931 under the command of Captain M. G. S. Hopson. From Port Said the company went by train to Jerusalem and Haifa. Headquarters, workshops and half the company were stationed in Jerusalem, where the Force Headquarters and one battalion were located, and the other half, under Lieutenant Kekewich, with ten vehicles, was stationed at Haifa with the other battalion. Until 14 Company arrived, the force had depended for its mobility on hired civilian transport. The first task of the company was to take over this miscellaneous collection of transport, and thereafter to be entirely responsible for the operational mobility of the force. Initially the strength of the company in Jerusalem was fifty-six, and of the detachment at Haifa, thirty-one. In the event of operations, the company was still dependent on requisitioned or hired transport to make it up to the required strength of vehicles.

The late 1920s appeared comparatively tranquil. Events in Europe had not yet stimulated emigration and the forces of law and order seemed adequate. In addition to the Palestine Police and the two Infantry battalions, the Royal Air Force had one and a half bomber squadrons and an armoured car company with HQ RAF in Jerusalem, and the Trans-Jordan Frontier Force, an efficient force of Arab levies with British officers, helped to maintain the peace on the border between Palestine and Trans-Jordan.

The part played by the Royal Army Service Corps was a most important one. Every emergency scheme for dealing with civil disturbance depended on mobility. To make all the available troops 100 per cent mobile entailed a five-fold expansion of RASC transport by hiring or requisitioning. This was twice necessary within a period of four months in 1933.

The activities of Hitler and the Nazis let loose a flood of immigrants in 1932, which caused violent Arab protests. In 1933, 30,000 Jewish immigrants were officially recorded (with many more irregulars, which were unrecorded); in 1934 there were 42,000 and in 1935 the figure rose to 62,000. During the period the amount invested in land purchase was £3,000,000 a year. In 1922 the alien minority amounted to a ninth of the population. By 1936 it had risen to a fifth.

In 1936, in the face of Arab protests, a plan for an elected Legislative Council was proposed, but as this would mean dominance by the Arab majority it was opposed by the friends of Zionism in Parliament. The disillusioned Arabs broke out in open revolt and organised a military force under Fawzi-ed-Din, a Syrian soldier of fortune.

In the previous fifteen years, disorders and civil disturbances had been quite common occurrences, but they were mainly Arab riots directed against the Jewish settlers. The main task of the forces of law and order had been to protect the scattered Jewish communities. But now the para-military forces of both Arabs and Jews were to be directed against British rule.

Early in 1936 the need for reinforcements became urgent. A light tank squadron was hurriedly sent from Egypt, and in September the 1st Division was despatched from Aldershot. It was a fortunate coincidence that the Italian conquest of Ethiopia was completed in May of that year, and the end of the emergency in Egypt made it possible to transfer RASC units from Cairo and the Western Desert to Palestine, instead of sending the Divisional RASC from home.

At the end of May 1936, 17 Company, in Egypt since the previous November, entrained for Palestine under the command of Captain G. K. S. Hewett, and was sent to Haifa, where it relieved the detachment of 14 Company, which rejoined the rest of the company in Jerusalem. 17 Company was at one time 260 strong and had detachments at Nazareth, Beisan, Rosh Pina, Safad, Jenin, Tulkarm and Nablus, covering most of the hilly country in northern Palestine.

19 Company left Mersa Matruh on 13 July after eight months in the desert and, after being re-equipped with 111 vehicles in Abbassia, arrived in Jerusalem a week later. There the company's workshop personnel were transferred to 14 Company in exchange for that company's drivers, and 14 Company took on the role of a mobile repair unit for the combined companies.

The Royal Army Service Corps component of the British Emergency Force in Palestine was completed by the arrival of 5 and 49 Companies during the next two months. In November 1936, 5 and 17 Companies were under the command of CRASC 5th Division (Major P. A. Arden OBE) in northern Palestine, and 14 and 49 Companies, at Jerusalem and Sarafand, were commanded by the CRASC 1st Division (Major E. B. Rowcroft MBE, who was made a temporary lieutenant colonel on 21 December 1936). 19 Company and ten other officers on temporary duty came under the Force Headquarters (ADS&T, Colonel G. A. Campbell DSO).

Martial law had been declared in April 1936 and the Palestine Emergency developed into a regular campaign, which lasted two years, in which British Forces suffered 300 casualties. The nature of the country favoured guerilla activity. Narrow mountain roads with hairpin bends, reckless native drivers and mules, donkeys and camels, were perpetual hazards. All movement had to be organised in armed convoys and escorts were the rule for individual vehicles. It was necessary to maintain a constant alert for snipers and ambushes. Lieutenant J. M. High (who by a strange quirk of coincidence always appeared next to Lieutenant Low in the Army List) was wounded when he was in charge of a 17 Company convoy. Driver Batey managed to drive the vehicle back to safety, with the officer slumped over the steering wheel. Driver J. K. Knott of 19 Company was awarded the Military Medal in the first list of awards for the emergency operations in Palestine, between 15 April and 14 September 1936. Before the end of the year two other awards of the Military Medal were made to RASC drivers. Driver J. G. Frank of 19 Company received his medal for conspicuous gallantry near Hebron, when driving an ambulance; and Driver J. H. Kent of 17 Company for conspicuous gallantry as a lorry driver, when he rescued a wounded soldier under fire near Ar Rama, during an action between 'D' Company, 1st Battalion The York and Lancaster Regiment, and a party of Arabs. During the same period ten officers and fifteen other ranks were mentioned in despatches, and in the following year, in the Coronation Honours List, MSM R. A. Bray and CSM W. E. Pope were awarded the MBE for valuable services in Palestine.

By December 1936 the force was reorganised and 5, 19 and 49 Companies returned to Abbassia in time for Christmas, leaving 17 Company in northern Palestine and 14 Company in the south.

49 and 51 Companies and most of the personnel of 19 Company sailed for home in the *Neuralia* from Alexandria, and arrived at Southampton on 3 February 1937. Headquarters RASC 1st Division also arrived home before March 1937. During 1936 the troops in Palestine increased seven-fold and by the end of the year there were twenty-three infantry battalions and one mechanised cavalry regiment in the country. To meet the situation a Main Supply Depôt was established at Haifa, with Supply Depôts at Haifa, Jerusalem and Sarafand, where the Army and Royal Air Force depôts had been amalgamated into a joint services depôt.

Before 19 Company left Palestine, the Commander of the British Forces in Palestine and Trans-Jordan, Lieutenant General J. G. Dill CB CMG DSO (later Sir John Dill, Chief of the

Imperial General Staff) sent the company a special farewell message, including the words: 'I have heard much of the good work you performed in the Western Desert under really bad conditions and how cheerfully you did it. Since your arrival in Palestine in last July you have carried out tactical lifting, embussing and general transport duties under what have amounted to active service conditions . . . Your record in Palestine is a proud one, and will take a very worthy place in the annals of your Royal Corps.'

At a lower level, the Commanding Officer of the 2nd Battalion The Lincolnshire Regiment, wrote to the OC RASC in Jerusalem on 4 December 1936:

'Before leaving Palestine, I wish to let you know of the most satisfactory work which the different MT detachments from 5 and 19 Companies have performed during their time at Tulkarm.

'We have demanded much from them, and their duties have entailed long hours, but it has been a pleasure to have had such willing and cheerful men. It has also been a source of great satisfaction to me to realise that I could always depend on them under all conditions, and particularly when employed in convoy work with my columns, on which occasions coolness and determination were required, as the roads passed along were, in several instances, under the fire of numerous snipers.

'I think that you would also like to know that the conduct and bearing of all were entirely above reproach.

'I should particularly like to mention the good work of T/36529 Cpl J. Robinson who was here for several months. His efficiency and method did much to ensure the smooth working and valuable co-operation to which I have referred.'

19 Company remained in Egypt for another year. The company was given a troop-carrying role, which it carried out until the formation of 67 Company in November 1937. The new company absorbed most of the personnel of 19 Company, which returned home in December and was re-formed at its old home at Catterick.

To try to break the deadlock between the Jews and Arabs, a Royal Commission was appointed, with Lord Peel at its head, which recommended partition of the country between Jews and Arabs, with a reserved neutral area enclosing the Holy Places in Jerusalem. This was acceptable to neither side. The deadlock continued for two years, the Arabs continuing their desultory campaign against the British until, under the influence of neighbouring Arab rulers, a more moderate attitude prevailed, and hostilities came to a temporary end in 1938, although sporadic outbursts of violence recurred.

During these two years the number of British troops in Palestine steadily increased, not necessarily because they were needed for operational purposes in Palestine, but because the Anglo-Egyptian Treaty of 1936 had limited the number of British troops in Egypt, and Palestine provided an alternative base for a strategic reserve in the Middle East.

The withdrawal of troops in 1937 had left a reduced garrison of only six battalions. This reflected an unjustifiable optimism, and the force was once more built up until, in 1938, there were some 23,000 troops deployed in addition to the Royal Air Force and the Police.

In Haifa and the northern area there was an increase of three infantry brigades and two cavalry regiments, a general hospital and a casualty clearing station, and a new divisional headquarters — 8th Division — was formed.

The formation of seventeen new Royal Army Service Corps companies, which had been authorised early in 1938, made it possible to send additional units from England. 61 Company, which had been formed at Catterick, embarked at Southampton at the end of September 1938 and arrived at Haifa in the middle of October, equipped with new Bedford lorries.

58 Company, under the command of Major H. K. Barnes, sailed in the *Aquitania* and joined 14 Company, which had moved from Jerusalem, and 67 Company, which had come from Egypt on a temporary assignment, at Sarafand.

68 Company, with an establishment of five officers and 142 other ranks, was specially formed for service in Palestine in September 1938 and went to Jerusalem, with a detachment at Sarafand. Headquarters RASC 7th Division (Major C. H. Massé MC) arrived at about the same time.

All these increases brought the number of RASC companies to six by the beginning of 1939, deployed as follows: in the Northern Area, with headquarters at Haifa (OC RASC, Major P. A. Arden OBE), there was CRASC 8th Division (Major J. A. Middleton) with 17 and 61 Companies. In the Southern Area (OC RASC, Lieutenant Colonel C. E. Davis) there was CRASC 7th Division (Major C. H. Massé MC) with 14, 58 and 67 Companies in Sarafand and 68 Company in Jerusalem.

The Royal Army Service Corps strength in Jerusalem was 11 officers and 219 other ranks, distributed between the staffs of Force Headquarters, HQ 7th Division and HQ 18th Infantry Brigade, 68 Company and the Supply Depôt. From the end of 1938 the Combined Supply Depôt at Sarafand ceased to be a joint Army/Royal Air Force unit and became the RASC Supply Depôt, responsible for supplying both the Army and the Royal Air Force.

Although conditions were more settled, the role of the Corps continued to be a war one. A full scale of rations was delivered to all units engaged on operations and all ration convoys were responsible for their own protection.

A fourth award of the Military Medal went to Driver V. S. Molin of 14 Company for gallant and distinguished conduct in Palestine. He 'showed conspicuous gallantry in action near Nablus on December 14 1938'. Several more officers, NCOs and drivers were mentioned for gallantry in General Orders.

By the middle of 1939, as war clouds gathered, 67 Company had returned to Cairo and, shortly after, 58 Company followed them and found themselves in tented accommodation at Mena Camp, beside the Pyramids, on the edge of the desert.

In 1939 Malcolm MacDonald, the Colonial Secretary, made concessions to the Arabs by placing a limit on Jewish immigration of 75,000 during the next five years, during which a new constitution would be worked out, with the object of granting independence to Palestine. This naturally aroused strong Jewish antipathy and the British were once again faced with active opposition from both Jews and Arabs. The outbreak of war in September 1939 set a moratorium on further action and both sides tacitly agreed to wait until the end of the war before continuing the struggle. The British mandate in Palestine was ended on 14 May 1948 and, on the following day, the state of Israel came into existence.

ABOVE: Move of 5 Company RASC to Palestine, leaving Alexandria on 8 September, 1936. CENTRE: On the Alexandria-Cairo road, a mixture of oil and sand laid by Shell Mex Co Ltd. BELOW: 19 Company RASC preparing a cricket pitch in Palestine, using an oil and sand mixture. Note the flimsies for the carriage of the oil, and the fly swat.

LEFT: The GSM with the Palestine Bar. BELOW: The 3-ton 6-wheeler Thornycroft, with its powerful 6-cylinder engine was extensively used by RASC units during the Palestine Emergency. It had an excellent performance on all types of terrain.

Part III
The Life and Structure of The RASC

Inside the War Office XIV

The War Office had greatly expanded during the 1914-18 War. When peace was declared, the Department of the Quartermaster General to the Forces reverted to its pre-war pattern, with Royal Army Service Corps affairs dealt with by the branches QMG 3 (Road Transport), QMG 5 (Royal Army Service Corps Personnel, Training and Organisation) and QMG 6 (Supplies). These three branches were staffed by Royal Army Service Corps officers. Each had at its head a senior officer (of the rank of colonel after 1923) with the title of Assistant Director of Supplies and Transport (ADST), and two other Royal Army Service Corps staff officers, a Deputy Assistant Director of Supplies and Transport (DADST) and a staff captain. These branch staff officers were directly responsible to the Quartermaster General, but also owed allegiance to the Director of Supplies and Transport, who was not provided with a separate staff of his own.

Until he retired in 1919 Major General Sir Frederick W. B. Landon KCMG CB was Chief Inspector of Quartermaster General's Services, which included the Royal Army Service Corps. When he retired, no Inspector of RASC Services was appointed until 1923. Instead there was a deputy under the Director of Supplies and Transport — Lieutenant Colonel H. N. Foster, followed by Colonel F. H. Brooke and Colonel G. F. Davies. When the appointment of Inspector of RASC Services was revived in 1923, Colonel Davies was appointed, with the title Colonel on the Staff, the equivalent of the rank of brigadier general, which had been abolished in 1921. At the same time the appointment of Deputy Director of Supplies and Transport was allowed to lapse. As Inspector he was directly responsible to the Quartermaster General. This arrangement continued until 1931 when the responsibilities of the Inspector were altered.

The Corps was also closely concerned with other branches of the Quartermaster General's staff. QMG 1 dealt with quartering, QMG 2 with movements and QMG 4 with veterinary services. From time to time Royal Army Service Corps officers held staff appointments in these branches, notably Colonel G. E. Badcock as AQMG and Major H. M. Gale as DAQMG in QM 2.

In 1927, under a reorganisation of the departments of the Quartermaster General and the Master General of the Ordnance, the former became responsible for the Inspectorate of Army Ordnance Services and the Directorate of Equipment and Ordnance Stores (DEOS), and four new branches were formed, QMG 7, QMG 8, QMG 9 and QMG 10, to deal with the affairs of the Royal Army Ordnance Corps. In addition QMG 11 was formed to be responsible for regimental institutes, regimental funds and miscellaneous services.

This arrangement was short lived. In 1930 the whole of the Ordnance Services reverted to the Master General of the Ordnance and the Directorate of Works and Fortifications became the responsibility of the Quartermaster General, with a reallocation of duties to the branches QMG 7, QMG 8, QMG 9, QMG 10, QMG 11 and QMG 12 and with a new branch, QMG 13, becoming responsible for technical transportation matters.

The Ordnance Services remained divorced from the Quartermaster General's Department until after the outbreak of war in 1939. In 1937 the appointment of Director General of Munitions Production was created and the holder became a member of the Army Council. Shortly afterwards, in February 1938, the four military members, who had been part of the Army Council since its constitution on 6 February 1904, were reduced to three, with the removal of the Master General of the Ordnance, whose office was abolished. A Deputy Master General of the Ordnance was retained under the Director General of Munitions Production, who took over all the responsibilities of the MGO, including the three Directorates of Armaments, Mechanisation and Ordnance Services.

In the early 1920s the head of QMG3 (Transport) was Lieutenant Colonel C. W. Macleod, followed by Colonel J. M. Young and Colonel B. L. Beddy. Colonel J. C. Browne, who later became Brigadier i/c Administration in Egypt, was at QMG5 and Colonels W. M. Parker and W. K. Tarver at QMG6.

There was no change in the organisation of supplies and transport until General Gibb became Director of Supplies and Transport on 9 August 1929; four months later, in January 1930, Colonel D. C. Cameron was appointed Assistant Adjutant General AG8/Assistant Director of Supplies and Transport QMG5, on completion of his tour as Commandant of the RASC Training Centre. This new dual appointment, in which Colonel Cameron was responsible to the Adjutant General for the administration of RASC personnel, and to the Quartermaster General for civilian personnel in the Corps establishments, represented an important step forward; henceforth the Adjutant General would be directly involved in the appointments of all Corps personnel.

An even more important change took place in responsibility for training of the Corps, which until then had lain with the Quartermaster General's Department. When he was at Aldershot, Colonel Cameron soon saw that, unless the General Staff assumed this responsibility, training of RASC units would always suffer in the competition with day-to-day administration. For ten years Corps unit commanders had fought a losing battle, trying to extract agreement to the hiring of civilian transport, in order to free their companies for essential war training. Now that he was head of the very branch concerned with RASC personnel and training, Colonel Cameron exerted every effort towards securing one of the most important decisions affecting the efficiency of the Corps.

Before the end of the year, the War Office announced that RASC training for war would be dealt with by the General Staff. Supervision of training at the Corps Training Centre would be the responsibility of the Director of Military Training.

An immediate result was the appointment of a GSO2 to the establishment of the Training Centre, with effect from 1 April 1931. The first officer to fill this appointment was Major G. Lammie MC, the Seaforth Highlanders.

At the same time (1931) it was laid down that the Inspector of RASC Services (Brigadier C. W. Macleod — the title of brigadier, in place of brigadier general had by now been restored, but only as an appointment) would report to the Chief of the Imperial General Staff (through the Director of Military Training) regarding the state of training for war of the Corps and to the Quartermaster General on other matters. The rank of the appointment continued to be brigadier until 1939, when Brigadier R. T. Snowden-Smith was promoted to major general.

From now on a Corps staff officer was appointed to the staff of the Director of Military Training. The appointment was that of a general staff officer, second grade, at first in MT1, then MT2 and MT3, and finally (in 1943) MT7. In addition to his responsibilities to the Director of Military Training, he also acted as staff officer to the Inspector of RASC Services and accompanied him on most of his visits of inspection. The first officer to be appointed (in 1931) was Major H. H. Dempsey; he was succeeded by Major H. N. Gallagher, Major P. S. Whitcombe and (in 1939) Major R. T. Cooke.

And so the Royal Army Service Corps had finally reached the position where the three leading military members of the Army Council were directly involved in the efficient organisation, administration and training of the Corps for its role in war.

Colonel Cameron was regarded by his contemporaries as one of the most outstanding Corps officers and he retired on 24 May 1932. After his death in 1955, three senior retired officers (two of whom had been Directors of Supplies and Transport) wrote a monograph in appreciation of his outstanding contribution to the Corps. In some ways his career presented a close parallel to that of Colonel Sir H. F. P. Percival. Neither reached the highest ranks in the Corps and yet they contributed more than most others to the improvement of the organisation of the RASC and the enhancement of its good name throughout the Army.

Thus General Gibb's tour as Director of Supplies and Transport witnessed the most important changes in the organisation of the Corps during the twenty years under review.

In 1934 the idea of a co-ordinating branch in the QMG's Department was born and QMG1 (co-ord) was created, to co-ordinate the work of the various QMG branches. Colonel G. E. Badcock, who had just finished a tour of duty as an instructor at the Senior Officers' School at Sheerness, was given the appointment of AQMG.

Other RASC War Office appointments were those of ADMT (Assistant Director of Military Transport), who had his office at Woolwich and was responsible for the War Department Fleet, the CIMT (Chief Inspector of Mechanical Transport) and his staff of inspectors, in whose office Colonel A. D. Owen spent the greater part of his career as an inspector of mechanical transport, until his retirement as Chief Inspector in 1934; the CIST (Chief Inspector of Subsidised Transport); and the Royal Army Service Corps representatives on such bodies as the Mechanisation Board.

In 1940, to meet the requirements of a greatly expanding organisation, a radical change took place. The Director of Supplies and Transport was given his own staff branches, with the initial letters ST, corresponding to the QMG branches. ST3 became the new staff branch dealing with transport in place of QMG3, and ST6 took over the responsibility for supplies from QMG6. Because of its growing importance, a new branch, ST2, was formed to deal with petroleum products. ST4 was responsible for Army catering and ST5 for the administration of civilians employed in Royal Army Service Corps establishments. ST1 was formed later in the year to take over the work of the Assistant Director of Military Transport, administering the War Department Fleet and all other aspects of water transport. This new concept did not involve an immediate doubling of staffs, as the existing officers were, for the most part, simply transferred from the QMG to the ST branches.

There is no doubt that the Supplies and Transport Directorate during the 1939-45 War measured up most efficiently to the enormous tasks placed upon it, but it has often been suggested that the creation of parallel staffs in the Q branches, and in the directorates, inevitably led to duplication and, in some cases, to misunderstandings.

A previous QMG, General Sir Redvers Buller VC GCB GCMG PC, was responsible for the reorganisation of the ASC.

LEFT: Major General Sir Frederick W. B. Landon KCMG CB, Chief Inspector of QMG Services, 1919.
RIGHT: Lieutenant General Sir Humfrey Gale KBE CB CVO MC, in 1929, DAQMG in QMG 2 (Movements) War Office; during World War 2 he was Chief Administrative Officer to General Eisenhower.

The Divisional RASC XV

In 1918 the Divisional Train consisted of four horsed transport companies, whose function was to maintain all the units in the division (including attached units) with supplies and forage. In addition to the daily ration, 'supplies' included commodities such as fuel, Royal Engineer stores, Ordnance stores and mail. The amount of petrol required in the divisional area was almost negligible. The Divisional Ammunition Column (DAC) was a Royal Artillery unit, which carried out the function of second-line transport forward of the ammunition refilling point. It consisted of three sections, two for gun ammunition and one for small arms ammunition and grenades. This last section made deliveries of small arms ammunition to the three Infantry brigades at ammunition points. The Divisional Ammunition Column was replenished from the Divisional MT Company ASC at refilling points.

The first change in this organisation was made at the beginning of 1921, when the supply of small arms ammunition, grenades and pyrotechnics within divisional areas was transferred from the Royal Artillery to the Royal Army Service Corps, a fifth company being added to the HT Divisional Train.

The development of the 30-cwt lorry with pneumatic tyres, pioneered in the experimental workshops of the RASC Training Establishment at Aldershot, convinced the authorities that the time had come to 'mechanicalise' the Divisional Train. In the revised edition of Field Service Regulations Vol I, issued in June 1923, the new organisation (approved in 1922) of a Divisional Train consisted of a train headquarters and two mechanical transport companies and one horsed transport company. No 1 (MT) Company was for the carriage of SAA and baggage and miscellaneous stores; No 2 (MT) Company was responsible for supplies. Each company consisted of a headquarters, workshops and four sections, one for divisional troops and one each for the three Infantry brigades. No 3 (HT) Company constituted a reserve, capable of carrying the rations and forage for three Infantry brigade headquarters and twelve Infantry battalions. At the same time a 'double echelon' system, as an alternative method of supply, was inaugurated. There was no headquarters divisional RASC.

This was the organisation which existed when the first post-war collective training exercises were carried out in 1924.

During the first five years after the end of the war, the opportunities for collective training were restricted, owing to the need to meet abnormal commitments with reduced resources. The emphasis had to be on individual training and field training at unit and brigade level. Consequently there was little time to study administration in the field, or to test the efficiency of new organisations, which were evolving with the rapid development of mechanical vehicles and the increased mobility which they conferred on commanding officers.

One of the main purposes of the 1924 manoeuvres was to put this newly won mobility to the test and to study the administrative problems involved, with special emphasis on traffic control, defence against air attack, the problems of maintenance and ammunition supply, the organisation and employment of the newly 'mechanicalised' trains, and petrol distribution in front of railhead.

During these exercises a new system of ammunition supply was tried out. The Divisional Ammunition Column (RA), with its mixture of horsed and mechanical transport, had been regarded as a somewhat unbalanced unit, and its theoretical radius of action from the refilling point was only seven and a half miles. The radius of action of the Divisional Train from refilling point was twenty-five miles. Similarly, the link between railhead and refilling point, covered by the Divisional MT Company (later called the Divisional Maintenance Company), was also twenty-five miles, giving a total range of fifty miles in front of railhead.

An improvised ammunition company of the Divisional Train was formed, equipped with 30-cwt lorries, to take the place of (a) 'B' Echelon (MT) of the Divisional Ammunition Column and (b) the ammunition lorries (SAA and grenades) of No 1 Company of the Divisional Train.

This experiment proved most successful under exercise conditions, paving the way for the eventual transfer of second–line ammunition supply from the Royal Artillery to the Royal Army Service Corps in 1929.

Another important lesson was the need for a much closer relationship between the commander of the Divisional Train and divisional headquarters.

The Times special correspondent summarised the manoeuvres under the heading 'Triumph of Mechanical Transport'. The correspondent of *The Commercial Motor*, who followed the exercises, was most impressed by the performance of 'these pneumatic tyred 30-cwt lorries and these four wheel driven pneumatic tyred tractors . . . the groundwork . . . of a new, interesting and highly efficient MT organisation', and summed up his impressions with the words: 'Indeed, it may be said that, in this new departure, the usual order of things is being reversed. At one time it was considered, if not exactly right and proper at least a thing to be expected, that civil enterprise should give, in matters of this kind, a lead to the War Department, as indeed to the majority of Government departments. In the present instance the War Office is showing the way and, by its adoption of a light, pneumatic tyred and, therefore, speedy vehicle, is in advance of public opinion to a remarkable degree.'

As a result of the 1924 manoeuvres a new establishment was proposed for the organisation of the RASC in the field, and this was tried out during collective training in 1925.

Under this a Headquarters Divisional RASC was included. The organisation of the Divisional Train reverted to having three Infantry brigade and one divisional troops composite companies. A separate ammunition company for the carriage of small arms ammunition, the Reserve HT Company and an innovation, a divisional mobile repair unit, were provided. This last unit, which was designed to deal with all second line repairs in the division, was formed to conform to a recently promulgated principle that no field unit should bear on its establishment its own workshop. It was several years before this principle was reversed.

During the manoeuvres it was found necessary to modify the experimental organisation, and in almost every formation taking part the final form of the Divisional RASC consisted of a baggage column (which also carried petrol), a supply column and an ammunition column, mobile workshops and the horse transport company (imaginary). The importance of brigade liaison officers (RASC) and the need for better means of communication within the Divisional RASC was fully demonstrated.

For the next three years the arguments for and against commodity companies or composite Infantry brigade companies went on, with the commodity company gradually gaining ground. In the next large scale manoeuvres in 1928 a CRASC (Commander Royal Army Service Corps) and an adjutant were appointed for each division taking part, and a divisional supply company, a divisional baggage company, a divisional 'bus company (for troop carrying) and a divisional repair unit formed the Divisional RASC. There was no horsed transport unit and the handling of petrol was hardly practised. Considerable use was made of civilian hired vehicles and drivers, as in all previous large scale exercises.

Largely as a result of the experience gained from these manoeuvres a new organisation for the Divisional RASC in war was approved in 1929. In place of the Divisional Train there was to be a Headquarters RASC, consisting of a CRASC, senior supply officer, adjutant and two officers for requisitioning duties, and three self-contained companies: the divisional ammunition company, the divisional supply company and the divisional baggage company, each with its mobile workshops and artificers. At last the Divisional RASC was fully mechanised, with adequate representation at divisional headquarters and with three companies designed for their particular roles. By early 1932 the new organisation was adopted for the Territorial Army divisional trains, bringing them in line with their regular counterparts. At the same time the Divisional Ammunition Column (Royal Artillery) was abolished, and the Royal Army Service Corps became entirely responsible for the carriage and delivery of ammunition from railhead to the wagon lines of Royal Artillery units. The divisional supply company continued to be responsible for the delivery of petrol.

The economic depression of the early 1930s was not allowed to interfere substantially with the progress of mechanisation in the Army. Since the 30-cwt 6-wheeled lorry had been adopted as the main equipment of the Divisional RASC, the number of vehicles on the war establishment of an Infantry division had been steadily increasing. Senior commanders had become conscious of the danger of air attack on concentrations of transport, as the amount of transport within the divisional area increased. In 1931 the GOC of the 1st Division at Aldershot, Major General Harman, who earlier in his career had served for eleven years in the Corps, was determined to reduce the number of vehicles in a Division. In that year's divisional training he exercised his division at full war establishment of transport, to test the efficiency of the divisional organisation. He planned, as far as possible, to replace 30-cwt lorries with 3-ton 6-wheeled vehicles. Development of the 3–tonner had reached a point where its cross-country performance could compare favourably with the 30-cwt vehicle.

In the same year a great step forward was taken by the decision to organise RASC companies in peacetime as divisional companies (albeit in cadre form) with their allotted war roles. By the end of the year, Cs RASC had been appointed for the 1st and 2nd Divisions at Aldershot, and in March of the following year divisional supply companies, baggage companies and ammunition companies had been formed for the 1st and 2nd Divisions. On 1 April 1933 the Divisional RASC of the 3rd and 4th Divisions were formed.

These divisional companies were mere skeletons of their war establishments, and on collective training had to be provided with additional vehicles from other units, or with hired civilian vehicles, many of which were of solid tyred types, even as late as 1935. For example, the 'standard' company, under the 1931 reorganisation of RASC MT units at home, had a peace establishment of two officers and 56 other ranks, with 15 MT vehicles and three motor-cycles.

But the most important achievement was that henceforth, the GOC of a division and his staff would be responsible for the discipline and training of the Divisional RASC units – the units with which they would go to war. This ensured a closer contact between the RASC and divisional staffs and a greater part in the field training of the division as a whole.

During the early 1930s the process of mechanisation began to be applied to Infantry battalions. Selected units were provided with a number of Carden Lloyd carriers and Commer pick-up trucks. Sir John Valentine Carden MBE had served in the Corps and when he died in February 1936 he was described as the 'foremost brain in the world of AFV (Armoured Fighting Vehicle) design'. Although the intention was to confer greater fighting mobility on the battalion, this development had two important effects on the Divisional RASC. The consumption of petrol within the division was considerably increased. In 1930 it was calculated that the daily requirement of petrol for an Infantry division was only 21 tons and that for a Cavalry Division 27 tons. By 1936 these quantities had more than trebled. The other effect was that the carrying capacity of unit first-line transport gradually went up. The result was that, when the next reorganisation took place in 1936, the carriage of petrol ceased to be a commitment of the Supply Company, which now became a Supply Column, operating on a double echelon system. The divisional baggage company became a divisional petrol company which, however, retained the responsibility for carrying blankets, anti-gas equipment and a reserve of clothing.

Thus, on the outbreak of war in 1939, the organisation of the Divisional RASC was:

Headquarters Divisional RASC.
Divisional Supply Column (Double Echelon).
Divisional Ammunition Company.
Divisional Petrol Company.

The organisation for a mobile division (later to become an armoured division) was modified to take account of the greater distances and wider dispersion involved. Second-line transport companies for each cavalry brigade and tank brigade and for mobile divisional troops were provided. These were composite companies, with sections for ammunition, petrol and supplies, the latter operating in double echelon. In addition, third line transport, consisting of a mobile divisional ammunition park, a mobile divisional petrol park and a mobile divisional reserve supply park, carrying three days reserve of preserved rations, formed an integral part of the mobile division.

Major General A. E. W. Harman CB DSO, Commander 1st Division, is with CRASC 1 Div and RASC officers in 1931 at the conclusion of the divisional transport exercise, held to evaluate the use of the 3-tonner in lieu of the 30cwt vehicle in the divisional RASC; top: Lt W. A. Winter, ADC Queens Bays; Capt W. Seaman RASC; Lt M. R. Briggs RASC; Capt C. B. Robertson OBE RASC; Capt R. P. Dent; bottom: Capt C. E. Davis RASC; Maj Gen A. E. W. Harman CB DSO, GOC 1 Div; Maj T. J. R. Langmaid MC CRASC 1 Div; Col C. J. Pickering CMG DSO, AA & QMG 1 Div; Capt L. A. N. Wright RASC.

Buller Officers' Mess in the 1920s.

Conditions of Service XVI

In the years between the wars, service in the Armed Forces offered few material rewards. Rates of pay were well below their civilian counterparts. Prospects of promotion were far from bright and, until the Hore-Belisha reforms of 1938, were often frustrating. In a number of Infantry regiments subalterns had to wait nineteen or twenty years for promotion to captain. Conditions were not so bad in the Cavalry, where the notice of impending foreign service for the regiment invariably brought a crop of resignations, or in the various Corps, where there were better opportunities for promotion to higher rank. In theory an officer had no right to resign his commission; an application to do so required the approval of the Army Council. The age for retirement depended on the rank reached by the individual officer, but only those who reached the rank of lieutenant colonel and above could serve beyond the age of fifty. Appointments were strictly controlled by establishments. If there was no vacancy in the appropriate rank an officer was placed on the half pay list until one occurred. A commanding officer's appointment was normally for four years, after which, if there was a vacancy in the establishment, he might be promoted to the rank of colonel but, if there was no appointment for which he was qualified, he was put on the half pay list until one became vacant. After a certain period, if no vacancy became available, he was placed on the retired list. Similarly, if an officer was in hospital or otherwise unfit beyond the stipulated period, he was placed on the half pay list until he was once more fit enough to be given a fresh appointment. In November 1933 there were six Royal Army Service Corps colonels on the half pay list and in July 1937 nine officers were on half pay.

A soldier initially engaged for a period of twelve years, of which only a portion was served with the Regular Army, the balance served with the Army Reserve. In Infantry regiments seven years were served with the Colours and five with the Reserve. In the Royal Army Service Corps the period of colour service depended on the man's trade and varied from time to time according to the Army's requirement for that particular trade. Normally a driver signed on for six years with the Colours and six with the Reserve, although in 1932, when recruiting was given a boost by the recession, it was modified for a time to three years with the Colours and nine with the Reserve. Most supply trades served three years with the Colours and nine with the Reserve, while artificers served seven years with the Colours and five with the Reserve. At the end of his colour service a soldier could elect to complete his twelve years with the Colours and thereafter re-engage for another nine years in order to qualify for a pension. A boy could enlist at the age of 14 and was required to serve for a number of years after reaching the age of 18.

One perquisite which the RASC officer enjoyed was 'Corps Pay', which recognised the special qualifications of Corps officers in the field of engineering. This had been withdrawn at the end of the war but was restored in the 1922 Pay Warrant for those officers who were qualified as AMIMechE (Associate Member of the Institute of Mechanical Engineers). In

April 1925 a lower rate of 'Corps Pay' was restored for those Corps officers who had satisfactorily passed the junior officers' course at the Training College, but had no engineering qualification. This rate started at 1s 4d per day for a subaltern and rose to 4s 8d for a lieutenant colonel.

In the same year command pay was authorised for Royal Army Service Corps officers who were promoted to lieutenant colonel.

However, these concessions were accompanied by reductions in the basic rates of Army pay. When new rates were laid down after the war it was explained that these were inflated by the high cost of living and were liable to reductions of up to 20% as the economy stabilised. Officers' pay was subject to revision every three years, according to the cost of living figures supplied by the Ministry of Labour. In October 1925 there was an overall reduction of 5½% and in February 1927 this was increased to 6%. In July 1930 a further cut of 2% was considered justified, which was spread over two years, the full 8% becoming effective on 1 July 1931.

Owing to the economic crisis a further cut of 3% in officers' pay, in line with the rest of the country, was introduced under the Economy Act in November 1931, with the proviso that no one was to suffer a greater cut than 10%.

Finally, a year later, it was decided to end this uncertain state of affairs and to stabilise officers' pay. It would no longer fluctuate according to the cost of living index figures. The final consolidation was deferred until 1 April 1934. Meanwhile the rates were fixed at 11% below the standard rates.

At the time, on commissioning, an officer received 9s 2d per day and a uniform allowance, which covered about 75% of his initial outlay.

The soldier's rate of pay in 1935 was 2s a day (all found) when he joined as a recruit. At the end of his training, if he had obtained a 2nd class certificate of education, he was entitled to an increment of 3d a day, and a further 3d after twelve months' efficient service. He earned an additional 6d a day after three years' service. From his 3rd to 12th month of service he received 3s 9d clothing allowance a month, which was subsequently increased to 11s 11½d a month.

Nevertheless, an Army career offered many attractions — an open-air life, plenty of opportunities of playing games and the chance of spending a fair proportion of one's life abroad, seeing the world at the Army's expense.

Although his rate of pay was so low, the soldier had few expenses and, provided he lived frugally, he could get by. He had the opportunity of learning a trade at no expense to himself and after six years' service as a driver or an artificer was well qualified for civilian life.

With more than half the Corps serving overseas it was almost certain that every soldier would serve part of his time abroad.

One of the attractions of service in the Royal Army Service Corps was the opportunity of serving in a great variety of home and overseas stations. In an Infantry regiment of two battalions, service was limited to the home battalion, the regimental depôt or the overseas battalion (which was most likely to be in India). In the Corps the choice lay between the training establishments and depôts and a large number of company stations and detachments both at home and abroad. The Royal Army Service Corps officer or driver might serve in any of the thirty-eight home stations or the thirty-six foreign stations where the Corps was represented in 1925. These included India (until 1928) and Rhine Army (until 1929), where the largest number of troops was stationed, Egypt, with garrisons in Cairo, Moascar (on the Canal), Port Said, Alexandria and Khartoum, Palestine, Cyprus, Malta and Gibraltar, the West Indies (Jamaica and Bermuda), Sierra Leone, Mauritius (the most southern Corps station, and the only one south of the Equator), Ceylon, Singapore, Malaya and Penang, and China (with garrisons in Hong Kong, Shanghai and Tientsin).

Conditions in these foreign stations varied greatly. Nearly all of them had their own particular attractions. The size of the Royal Army Service Corps component might be anything from a small detachment of two or three officers and warrant officers and a handful of NCOs and men, to a full-size lieutenant colonel's command. Three years in a station like Bermuda or Mauritius seemed to some to be quite idyllic, with ample time to enjoy the local amenities when the business of the day was done. Gibraltar, in spite of the confined area, offered many attractions to the keen horseman, with the local races and the neighbouring Calpe Hunt in Spain. It was, however, important that the members of such a small detachment were congenial and could bear each other's company for months at a stretch. For the more gregarious minded and the extrovert a station like Cairo offered an almost endless variety of amenities — excellent opportunities for realistic training in the desert, an extremely healthy climate (even in the hot summer months), interesting places of historical importance to visit on local leave and the chance to pursue any game or pastime at reasonable cost.

If the cost of living in an overseas station was higher than that in the United Kingdom, an overseas allowance was paid. For example, in Egypt in 1937 an unmarried subaltern received an allowance of 10s a day.

In the hot weather work started early in the morning — sometimes before breakfast — and in many overseas stations finished at 1.00 pm, leaving the rest of the day free.

Overseas service was governed by the Foreign Service Roster, which listed each rank in the order in which the officer or man was liable to foreign service. On return from overseas, or on promotion, an officer was placed at the bottom of the list for his rank. Arrangements could be made for the exchange of places on the roster with someone else lower down, if one's name was due to come to the top of the list at an inconvenient time, by the payment of an agreed sum.

Service in overseas stations was normally for a tour of three or five years (five years mainly for Mediterranean stations and India). In certain cases, for example in a station where families were not allowed, service was restricted to a year, the remainder of the tour being spent in another foreign station.

All movement to and from foreign stations was by troopship. The trooping season normally began in September and finished in May. Warning orders were issued for either the first half or the second half of the trooping season. Permission could be obtained to travel independently at one's own expense. The troopships — with the prefix HT (Hired Transport), or HMT (Hired Military Transport) — were operated by civilian shipping firms such as the British India Steam Navigation Company and the Bibby Line. Conditions on board were somewhat cramped, particularly for the soldiers, whose accommodation served the dual purpose of mess-decks during the day and, by slinging hammocks, of sleeping quarters by night. There were always sighs of relief if the crossing of the Bay of Biscay was a smooth one.

Military affairs on board were administered by the commanding officer of a major unit, if one was being moved, or by an officer commanding troops, who was specially appointed by the War Office from the officers being embarked, in the case of details. To assist him the OC troops would appoint a baggage officer, an entertainments officer and a messing officer from the passengers, for duty during the voyage.

One of the ship's staff was a 'troops officer', who was responsible for liaison with the OC troops on all matters affecting the military personnel on board.

The voyage to Egypt and the Middle East took from two to three weeks, and that to the Far East from five to six weeks. These schedules allowed ample time for acclimatisation, and soldiers arrived in a fit condition to take up their duties in their new station.

When HT *Dilwara* went into service with the British India Line in 1936 (11,300 tons) she was the first of a series of purpose-built troopships, which added considerably to the comfort of trooping.

The embarkation of troopships was controlled by the 'Q' Movements Branch of the War Office. Colonel G. E. Badcock CBE DSO held the appointment of AQMG, QMG 2, which was the responsible branch, from the beginning of 1934. He had a movement control staff at Southampton, the main port of embarkation, which consisted of a staff captain, Movements (any arm) and a RASC staff, headed by a staff sergeant-major, with two other warrant officers, five NCOs and a civilian staff of four. This staff was also responsible for embarkation at other ports, *eg*: Liverpool. Embarkation staff in overseas ports was provided by the Royal Army Service Corps. Captain L. W. Genders, who was ESO (embarkation staff officer) at Port Said, was well known to almost every ship's captain who passed through the Suez Canal in the early thirties. There were also MLOs (movement liaison officers) at Hong Kong and Shanghai, provided by the Royal Army Service Corps.

It was calculated that in seven years virtually the whole Army passed through Southampton. 1937-38 was a particularly busy trooping season, with more than thirty troopship sailings out of Southampton. In that year 1,645 officers, 744 wives and 432 children, and 24,082 other ranks, 2,070 wives and 2,697 children were embarked. 1,012 officers, 648 wives and 453 children, and 27,469 other ranks, 2,091 wives and 3,211 children were disembarked. These figures excluded those passing through Liverpool and those dealt with by the embarkation staff of the ADMT (Assistant Director of Military Transport, Woolwich).

One over-worked member of the movement control staff was heard to mutter the catch:
'Knock knock.
Who's there?
Isaac.
Isaac who?
I's sick of seeing troopships.'

During a tour of foreign service no home leave was granted at Government expense, but 'indulgence' passages could be allowed when spare accommodation was available in the troopships. If he was on a five-year tour it was normal practice for an officer to accumulate his leave entitlement (two months a year) and to take home leave about twice during the tour. Allowing for about a week's travel each way this gave him three and a half months at home. Most other ranks tended to spend the whole of their foreign tour abroad without home leave.

In addition to regimental service in Royal Army Service Corps overseas stations there were other opportunities for foreign service open to officers and men of the Corps. The Sudan Defence Force, the Supply and Transport Corps of the King's African Rifles, the Somaliland Camel Corps, the Royal West African Frontier Force and the Trans-Jordan Frontier Force all employed British officers and other ranks. Volunteers could apply for secondment to any of these forces, either in special appointments, mainly reserved for the Royal Army Service Corps, or for general service vacancies open to the whole Army. In all these forces Corps officers played an important part, the most significant achievement, perhaps, being that of Lieutenant Colonel C. V. Bennett (Brigadier C. V. Bennett OBE), who was given command of the Somaliland Camel Corps during his six years' service with the King's African Rifles (1932-38).

The Sudan Defence Force was the most 'prestigious' of these forces and the most popular with Corps officers. Throughout this period there were always between six and ten officers of the Corps on secondment.

Since the battle of Omdurman in 1898 the Sudan had been a condominium, ruled jointly by Great Britain and Egypt. The post-war troubles, culminating in the assassination on 19 November 1924 of General Sir Lee Stack, the Sirdar, who was Governor General of the Sudan and Commander in Chief of the Egyptian Army, led to the withdrawal of all Egyptian troops from the Anglo-Egyptian Sudan. In their place the Sudan Defence Force came into being in January 1925, under the administration of the Foreign Office, and was the direct successor of that part of the Egyptian Army which, prior to 1925, was permanently stationed in the Sudan. All the Arab and Sudanese units and the British officers serving with them were incorporated into the new Sudan Defence Force, which owed loyalty only to the Governor General of the Sudan. The force consisted of all arms and was organised into five areas — northern, southern, eastern, western and central — covering a large expanse of country. The force was self contained, with its own supply, transport, medical and veterinary services. Supply and transport were divided into three departments — supply, animal transport and mechanical transport — co-ordinated by the Assistant Quartermaster General at SDF headquarters. One Royal Army Service Corps officer was attached to the 'Q' branch. In 1929 there were two officers in the supply branch, four in animal transport and three in mechanical transport. Later the balance shifted in favour of mechanical transport (hamla mekanikia). By 1931 there were no officers left in the animal transport department. At that time there were six Royal Army Service Corps officers and fourteen NCOs seconded to the Sudan Defence Force. In theory these officers were interchangeable, although it was unlikely for a trained workshops officer to be moved from his specialist appointment.

The mechanical transport department had expanded from one company and workshops in 1928 (in north Khartoum) to three companies and five workshops, based on Khartoum North, El Obeid and Torit. The establishment of the workshops at Khartoum North had a complement of three British officers, two native officers, three British NCOs and a British civilian store-keeper. It was this unit which made a striking contribution to the war effort, by improvising the construction of an armoured car on an ordinary 30-cwt Ford chassis with reinforced springs, for use by the medium machine gun companies of the SDF. The vehicle was armour-plated and armed with a Vickers medium machine-gun and a Boyes anti-tank rifle mounted in a revolving turret. Some forty-five armoured cars of this model were produced and were successfully used against the Italians in Eritrea in 1940-41. The inspiration for this venture came from Miralai Stewart Bey (Brigadier C. G. Stewart CBE), who was in command of mechanical transport at the time. The workshops also produced an unarmoured Ford V8 Bren carrier which proved its worth against the Italians.

An applicant for service in the Sudan Defence Force was required to have at least five years service, to be unmarried and, if a subaltern, to have passed his promotion examination for Captain.

On appointment a Corps officer was given the rank of 'bimbashi' (equivalent to major), with pay of £720 per annum and generous travelling, servant, lodging and other allowances. He was entitled to three months leave every year, with an assisted passage allowance which was roughly equivalent to a single fare.

The initial contract, to serve under the Sudan Government in any part of the Sudan, was for two years. By mutual agreement this could be extended to five years, provided that a qualification in Arabic was obtained. The contract could be terminated by either side, with three months notice, unless the individual was under orders for active service.

A period of service with the Sudan Defence Force provided invaluable training and experience for a young Corps officer. At an early stage in his career he was given far greater responsibility than he could expect in any normal appointment. His pay and conditions of service were far better than those of a regimental officer. There was a great temptation for an officer to overstay his time with the force. Captain W. d'A. Collings (later Inspector Royal

Army Service Corps) was seconded to the Egyptian Army in December 1922 and was one of the original members of the Sudan Defence Force. He was not restored to the home establishment until 1 January 1931. Lieutenant Colonel C. O. Henfrey probably spent as long a time with the Sudan Defence Force as any other Corps officer (1932 to 1941). Other officers who served an apprenticeship in the Sudan and subsequently rose to high rank in the Royal Army Service Corps were Major General E. H. G. Lonsdale CB MBE (Inspector and Director), Brigadier E. R. Goode CBE (Inspector), Brigadier H. M. Gough CBE, Brigadier T. E. Southwell and Colonel N. F. Penruddocke (who was AQMG at Headquarters SDF).

Two Sudanese officers who served with the Royal Army Service Corps in the Sudan Defence Force from its original formation in 1925, must be mentioned. Saghalasi Ibrahim Abboud and Saghalasi Ahmed Muhommed served the SDF with the greatest loyalty and devotion and earned the friendship and respect of all the Royal Army Service Corps officers who served with them. When Sudan eventually won its independence, Abboud became the first President and Ahmed the first Sudanese Kaid el Amm.

Apart from the Sudan Defence Force there were other British troops stationed in the Sudan. There was a garrison of two infantry battalions, a Royal Artillery detachment and a military hospital. The Royal Army Service Corps detachment totalled three officers and about twenty soldiers, including the clerks employed at headquarters British troops in the Sudan, all found from British troops in Egypt. The British troops were all stationed in Khartoum, except for two companies detached from one of the infantry battalions — one at Atbara on the Nile, 150 miles north of Khartoum, and one at Gebeit in the Red Sea Hills about 400 miles from Khartoum and 80 miles from Port Sudan. Khartoum being a 'no family' station, the tour of duty was limited to one year, and normally formed part of a five-year tour in Egypt.

Of all the forms of extra regimental employment open to a Royal Army Service Corps officer, service on the staff was the most important. Officers of the Corps had had their share of important administrative staff appointments during the 1914-18 War. This was not surprising when it is considered that, in the field, more than 75 per cent of all 'Q' staff work is concerned with supplies or transport. By the nature of their training and experience they were more qualified to fill administrative appointments than officers from other arms and services. At the War Office the branches of QMG 3 and QMG 6 were concerned entirely with transport and supplies, and it was only natural that the staff for these branches should be filled by Corps officers. The extravagant and cumbersome organisation of two parallel 'Q' and 'ST' channels operating under the Quartermaster General had not yet been devised, and he and the Director of Supplies and Transport received identical advice direct from the experts on their staff, without the risk of dilution or distortion on the way, from crossed channels.

After the war some of these officers legitimised their position and obtained staff qualifications by attendance at the Staff College.

In the immediate post-war years a kind of reaction set in and there was a prejudice among certain Corps officers against staff training. Whether this was due to the restricted number of vacancies at the Staff College made available to the technical Corps, or whether it was because of a legacy from the war years, that somehow service on the staff for an active officer was slightly shameful, or whether the view prevailed that practical experience was more valuable than the academic and theoretical training of the Staff College, the fact remains that the excuse 'studying for the Staff College' was too often interpreted as merely a ruse to dodge playing, or watching games with the troops, or taking part in some other open-air occupation.

This was a short-sighted view and one which, fortunately for the Corps, was disregarded by a fair proportion of promising young officers, who were not carried away by the call of mechanisation and looked on the staff as a necessary step in any advancement to the higher ranks of the Army.

In spite of the fact that only one vacancy out of forty-three was allotted to the Royal Army Service Corps during the 1920s, there were usually at least two Corps officers attending courses at Camberley, and they all made their mark in the Corps in later years.

Although only a small proportion of the officers taking the Staff College examination could hope for a vacancy, the mere act of qualifying carried a distinct cachet and later, when the Army began to expand, many of those who had qualified found themselves in staff appointments, with the opportunity of earning advancement on the staff through their practical experience, instead of the academic instruction of a Staff College course.

There were other ways in which RASC officers could exercise their talents. Language courses gave opportunities for attachments to foreign armies and, in the case of Chinese and Japanese, provided almost a specialised career. Some officers were qualified as barristers — Colonel H. G. Reid, Colonel W. N. Craig-McFeeley, Major J. Reckitt and Captain E. D. M. Heriot-Hill. Others had a penchant for serving as military attachés. Captain H. J. Cooper was military attaché to the British Embassy in Teheran and Lieutenant Colonel C. S. Vale MC was military attaché in Riga.

The great variety of employment enjoyed by Royal Army Service Corps officers is well illustrated by this list of secondments in 1939. Apart from those holding appointments and attachments in the QMG branches of the War Office and in such establishments as the Inspectorate of Subsidised Transport, there were five officers on the staff at home and seven abroad, seven officers on the Inspection and Experimental Staff, Royal Arsenal, Woolwich, six officers at various army instructional establishments, including the Royal Military College, Sandhurst and the Military College of Science, one military attaché, two officers at British Missions to foreign armies, one officer on military exchange in Canada, seven officers with the Sudan Defence Force, three with the King's African Rifles, one with the Royal West African Frontier Force and one with the Palestine Police. In addition there was a Territorial adjutant seconded to each of the eighteen Territorial Army Formations. With all these opportunities, no one could with justice claim that life in the Royal Army Service Corps was dull or uninteresting.

There was, perhaps, some justification for the manpower branches being reluctant to let too many officers and men get dispersed into extra-regimental employment. During the first years of the peace, when the doctrine of 'no war for ten years' was accepted, the reduced size of the Corps was looked upon as inevitable, although the personnel branch was hard put to it to find officers and men for emergencies. When the decision to re arm was taken in 1934, mobilisation schemes at once resumed their importance in the scheme of things. The limited resources of such a small Corps to meet its minimum mobilisation commitments became immediately apparent. With less than 500 Regular officers (including quartermasters and mechanist officers) and under 5,400 warrant officers, NCOs and men, the Royal Army Service Corps could barely meet a third of its immediate mobilisation requirements. The position was worse than it looked, because at any time a considerable proportion of the strength was non-effective as far as Royal Army Service Corps employment was concerned. In June 1930 there were 564 other ranks on the non-regimental list, in a variety of appointments at home and abroad, who were beyond immediate recall — two of them with the Boundary Commission for Somaliland, Northern Rhodesia and the Belgian Congo. Of the 470 officers on the Regular establishment in 1931, after deduction of seconded officers, those in staff employment, those in key appointments, and a percentage for sickness and other casualties, it was calculated that less than 300 would be available to fill mobilisation appointments in Royal Army Service Corps units. The only possible conclusion was that great reliance would have to be placed on the Army Reserve, the Territorial Army and the Supplementary Reserve.

GROUPING FOR PAY, ROYAL ARMY SERVICE CORPS TRADES.

Group A.	Group B.	Group C.	Group D.	Group E.
		(a) PEACE TRADES.		
—	Wheeler	(i) *Animal Transport.* Farrier, saddler and harness maker.	—	—
—	Clerks (high speed shorthand writers)	(ii) *Supply.* Clerks, bakers, butchers	—	—
Draughtsmen (mechanical), electrician, fitter, turner	Blacksmith, carpenter and joiner, coach painter, coppersmith, moulder, welder (acetylene).	(iii) *Mechanical Transport.* Technical M.T. clerk, coach trimmer, storeman (Technical and Departmental), driver mechanic.	— —	— —
(b) IN ADDITION TO THE ABOVE TRADES, THE FOLLOWING WILL BE REQUIRED FOR WAR:—				
Millwright	Machinist (metal)	(i) *Mechanical Transport.* —	Vulcaniser	—

ABOVE: Grouping for pay RASC trades. BELOW: Forms of dress worn by RASC soldiers, in the UK in 1931, in front of Buller Officers' Mess; back L to R: HT Ceremonial (Dismounted); Trumpeter Walking out (Winter); HT Marching Order; Cook; Trumpeter Drill Order; MT & Supply Marching Order; Soldier in civilian clothes; Dismounted Walking out (Summer); Trumpeter Marching Order; Fatigue Dress; Baker; Drill Order Fatigue Dress; Dismounted Walking Out Order (Winter): PT Kit; centre: Bandsman Service Dress; Drill Order; Off Duty Dress Barrack Area; front: HT Walking Out Dress; HT Duty Order; Rough Rider; Bandsman Full Dress; PT Instructor; Blue Patrol; HT Full Dress Mounted; Ceremonial Order; Battle Order; Drill Order with rifle; Marching Order Fatigue dress; Marching Order Service Dress.

Kit Inspection UK in 1932, layout and RIGHT: detail.

Kit Inspection.—(Plate " C.")

)n the Man.

:ket, S.D.
ousers, S.D.
ots, ankle.
aces.
rt.
awers (cotton or woollen)
Socks.

Laid out for Inspection.

Name plate	... ...	(1)
Belt, buff	... ...	(2)
Walking-out cane	...	(3)
Equipment, complete		(4)
Coat, M.S. or greatcoat, D.S. (with titles) and second coat, M.S. underneath when issued.		(5)
Mess tin	... ...	(6)
Cover, mess tin...	...	(7)
Cap, S.D., stiff (with cap badge)	... ...	(8)
Jacket, S.D.	... ...	(9)
Trousers, S.D. ...	...	(10)
Cap, S.D., soft ...	...	(11)
Shirt	... ...	(12)
Drawers (cotton or woollen)		(13)
Socks	... ...	(14)
Puttees	... ...	(15)
Gloves, woollen	...	(16)
Gloves, driving (if issued)		(17)
Bible and/or Prayer Book		(18)
Army Book 64	...	(19)
Permanent pass.		
Medals (if in possession) on top.		
Holdall, containing :—		(20)
Razor, safety, with blade.		
Comb.		
Spoon.		
Fork.		
Knife.		
Brush, shaving.		
Brush, tooth.		
Brasses, cleaning.		
Knife, clasp and lanyard		(21)
Soap and boot blacking		(22)
Oil-bottle	... ...	(23)
Pull-through	... ...	(24)
Housewife	... ...	(25)
Cap-comforter ...	...	(26)
Waistcoat—cardigan ...		(27)
Towel ...	... ...	(28)
Brushes :—		(29)
Hair.		
Clothes.		
Polishing.		
Brass.		
Blacking.		
Bag, kit	... ...	(30)
Overalls (on bag, kit)		(31)
Boots, ankle (with laces)		(32)
Shoes, canvas (with laces)		(33)
Rifle (in rack) ...	...	(34)

At the Wash. (Normally).

Shirt.
Drawers (cotton or woollen).
Socks.
Towel.
Overalls.

NOTE.—(1) Both jackets, S.D., to have R.A.S.C. buttons.

(2) Any additional public clothing issued to a man will be shown on top of the overall suit laid out for inspection.

Barrack Room name plate holders.

(**117**) Brass name plate holders, with a moveable slip engraved with the soldier's name and number, as shown in Plate "A," will be used in all barrack rooms. The tongue of a regimental cap badge is put through the slot in the holder.

The holder will be Company, etc., property, but the brass name plate will be issued to a recruit on joining the Training Battalion, and will be taken by him when he moves from one station to another.

Colour of Blanco of web equipment.

(**118**) The colour of Blanco for use on web equipment and strappings of breeches of personnel of the Animal Transport Branch is " Pickering's Web Blanco No. 97."

THE WAR OFFICE,

17*th June*, **1932.**

A. A. MONTGOMERY-MASSINGBERD,
Adjutant General.

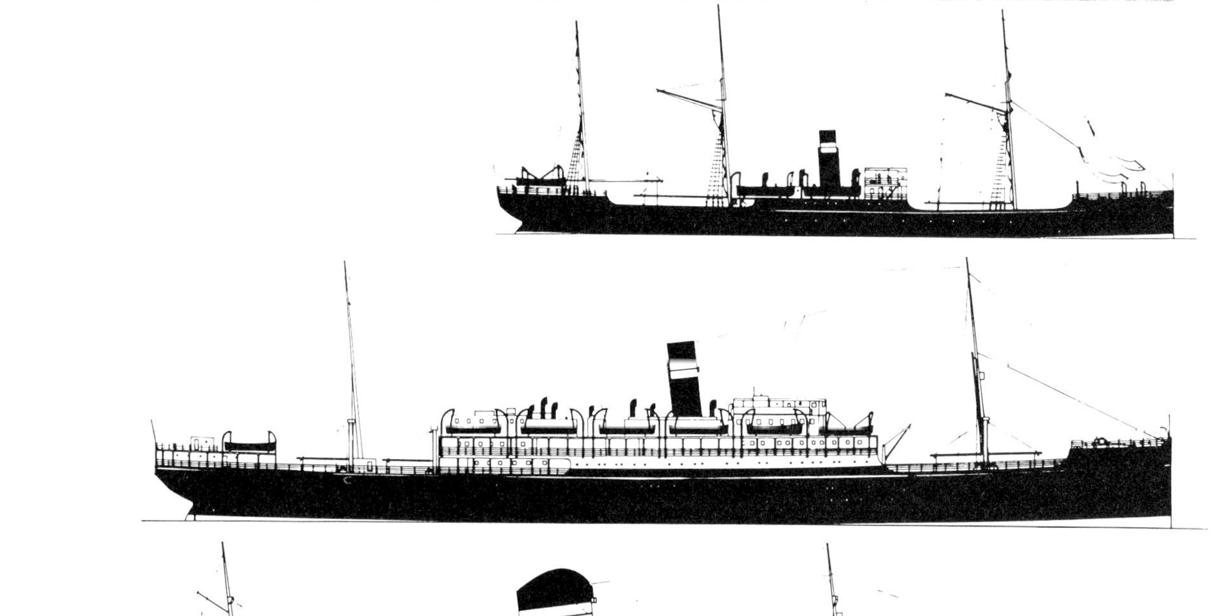

1884-1906

1913-1948

1956

OPPOSITE ABOVE: Parade Dress worn by RASC in Shanghai in 1928, and BELOW in Hong Kong in 1935. ABOVE: Marching Order Service Dress worn by RASC TA in 1936. BELOW: HT *Nevassa*, one of the first and one of the last troopships. (PGT)

TARIFF OF MEALS

BREAKFAST		1/-
LUNCH		2/-
DINNER		2/6
DINNER (Guest Nights)		3/-
TEA		6d.
EARLY MORNING TEA	*per cup*	2d.
COFFEE (Specially Ordered)	*per cup*	2d.
MESSING (Breakfast, Luncheon, Dinner)	*per day*	2/-

PRICE LIST

CHAMPAGNE.

	Vintage.	*bott.*	*half-bott.*
Louis Roederer	1911	**16/6**	**8/6**
Bollinger	1914	**16/6**	**8/6**
Perrier Jouet	1917	**13/6**	**7/-**
Duc de Prony	—	**9/9**	**5/-**

CLARET.

Medoc		**2/8**	**1/5**
Margaux		**5/-**	**2/9**
Grand Poujis Moulis	1916	**5/-**	**2/9**

BURGUNDY.

Volnay		**6/6**	**3/6**
Beaune		**4/9**	**2/6**
Gevrey Chambertain		**6/6**	**3/6**
Corton	1916	**7/6**	**4/-**
Pommard	1911	**8/-**	—
Pommard		**5/6**	—

HOCK & MOSELLE.

Liebfraumilch	**3/3**
Light Moselle	**2/-**
Zeltingen	**2/6**
Berncastel	**2/6**

WHITE WINES.

	bott.	*half-bott.*
Graves	**2/9**	**1/6**
Graves (Chateau Audene)	**5/3**	**2/10**
Barsac	**5/3**	**2/10**
Sauterne (Loupiac)	**5/3**	**2/10**

Royal Army Service Corps.

HEADQUARTERS MESS, ALDERSHOT.

Lieut J.W. Purves

CHEQUES SHOULD BE MADE PAYABLE TO "MESS SECRETARY, R.A.S.C., H.Q. MESS."

Mess Bill for	£	s.	d.
Messing at 2/- per day	2	14	-
Extra Messing		6	2
Tea at 6d.		11	-
Messing, Non-Dining Member			
Wines, Tobaccos, &c.	2	2	11
Maintenance for Month } Proportion 27/31		8	1
Monthly Subscription }		13	1
Sports Fund			
Band Fund			
Servants at 2/- per day	2	14	-
Billiards (Charge for Table)			
Telegrams and Telephones		1	2
Hire of Bedding, Carpets, etc.		8	6
Stamps		6	7
Cards			
Petty Cash Payments Polish & Soap 7d.			7
Breakages			9
Mess Guests			
Outstanding			
Total for Month	10	6	10
Bridge			
Billiards			
Credits			
RECEIVED £10-6-10 (cash). Mess Secretary. TOTAL £	10	6	10

Mess Secretary.
Royal Army Service Corps.

Payment of this account should be made by 7th inst.
(King's Regs. Para. 1458).

N.A.A.F.I. PRINTING.

Territorial Army
and Supplementary Reserve XVII

The Divisional Royal Army Service Corps (Territorial Army) had its origins in the Territorial Force which was created in 1908 by the Territorial and Reserve Forces Act of 1907. This was the outcome of a searching review of the whole structure of the British Army undertaken by Mr R. B. H. Haldane (later Lord Haldane), Secretary of State for War, with the able assistance and guidance of Colonel Sir Edward Ward, who was the Permanent Under-Secretary.

Before then there had been no officially formed supply and transport units in the Volunteers or Militia.

The new Territorial Force was designed to provide self-contained reserve formations complete with their own supporting arms and services.

The composition of the Territorial Force was defined thus: 'Instead of being . . . a fortuitous collection of units, representing but few of the arms and departments of an army, it will be organised upon the same basis as the regular force. The two forces, Regular and Territorial, will each consist of corresponding arms and departments and a community of interest will be thereby established.'

Under the new organisation the Regular Army of six divisions was to be supported by a Territorial Force consisting of fourteen divisions, each complete in itself, fourteen mounted brigades with their own horse artillery, transport and supply columns and cavalry ambulances, and other Army Troops. The primary role of the force was to be home defence.

Having been the Chief Supply Officer, under General Roberts in the South African War, Colonel Ward fully understood the importance of the Supply and Transport Service, and it was partly due to his influence that the Army Service Corps had such a good start in the Territorial Force. He gave it his full support and himself became the Honorary Colonel of the 2nd London Divisional Transport and Supply Column ASC (TF) in 1908, and was Chairman of the County of London Territorial Force Association.

There was no lack of enthusiasm in the recruitment of volunteers for the new transport and supply columns. The main problem was the general shortage of equipment and stores. Owing to the limited financial resources allotted to the Territorial Force, units were largely dependent on the co-operation of industry and the generosity of local land-owners and supporters for their resources. The hire of sufficient horses and wagons and one or two lorries for annual camp training was quite beyond the scope of the grants available to the County Associations, who were responsible for meeting these expenses. This was a problem which always had to be faced during the aftermath of a war and which was to be met again in the 1920s.

Nevertheless there was no shortage of enthusiasm. The keen spirit that was everywhere apparent was typified by the East Lancashire Divisional Transport and Supply Column, who

OPPOSITE LEFT: Buller Officers' Mess 1925 tariff and price list. RIGHT: An Officers' Mess bill, 1924. BELOW: Binbashi E. H. G. Lonsdale, whilst serving in the Sudan Defence Force; he later rose to become Director of Supplies & Transport, (1966-1969).

in 1910 took with them by road to camp on Salisbury Plain six motor wagons and trailers in addition to fourteen pair-horsed wagons and three one-horsed carts.

The fourteen transport and supply columns formed in 1908 largely retained their identity through two major wars, in which they amply justified the faith their supporters placed in them. Before 1914 the peace establishment of a divisional transport and supply column was 19 officers and 450 other ranks. A Territorial force adjutant and four permanent staff instructors were allotted to each column from the Regular Army Service Corps. The column was organised into a headquarters company and three (county) brigade companies (each with a supply section).

By 1913 every Army Service Corps unit had received training in all the transport and supply duties of their formations in camp and on manoeuvres.

When war was declared in August 1914 many units of the Territorial Force were in the middle of training in camp. Although it had not been envisaged that the Territorial Force would go overseas until six months after embodiment, there was an immediate call for volunteers for Imperial Service, in order to release Regular units for service on the Western Front. The call was met with an enthusiastic response.

The East Lancashire Division embarked for Egypt in September 1914 and 429 Company ASC was actively engaged at Gallipoli. The first Territorial Force divisions to go to the Western Front were the North Midland and London Divisions, who received their orders at the end of October 1914 and crossed to France in March 1915.

In December 1914 the War Office decided to number the Territorial divisions as part of the Regular Army. The numbers began at 42 and the fourteen divisions were numbered consecutively up to 55. For example, the 2nd London Division became the 47th London Division. At the same time the Divisional Transport and Supply Column was renamed the Divisional Train and the Divisional Supply Column. The Army Service Corps establishment was increased from 501 officers and men to 739.

The pre-war training of the Army Service Corps units of the Territorial Force was amply rewarded by the efficient way they carried out their manifold tasks. Before long no distinction was made between The Regular Army and The Territorial Force. The latter became an integral part of the British Army, although a certain pride in their former Territorial status was never lost.

At the end of the war, as the armies demobilised, there was no established procedure for the 'disembodiment' of the Territorial Forces, and they simply disintegrated. Those returning from the war kept the spirit and the name of their Territorial Force formations and units alive. They were ready to rebuild those units as soon as a decision was taken about their future.

Nearly two years elapsed before reorganisation was worked out and put into effect by the Territorial and Militia Act 1921. In March of the previous year the objectives of the reorganisation as the Territorial Army, were explained by the War Office:

'The present plan for the reconstruction of the Territorial Force is to adapt the old force to the military requirements of the future based on the experience of the war. The war has proved the high military value of the Territorial Force and the desirability of assigning to it a position of increased importance in the general organisation for Imperial Defence. Briefly, it is intended that the Force shall be reconstituted on a footing that will enable it to discharge to the full the role of a genuine second line to the Regular Army. It will therefore be organised on the lines of the Regular Army itself, with a view to supporting the latter in the event of a national emergency and to providing a means for any necessary expansion of the Forces of the Crown.

'Units of the Territorial Force, when embodied by Royal Proclamation and despatched abroad under authority of Act of Parliament, will not be used for the purpose of supplying drafts for the Regular Army. They will be used as regimental units and, if their efficiency warrants, as brigade and divisional units.'

The first line of the Regular Army was fixed at five divisions. The Territorial Army was composed of the same fourteen divisions which existed before the war, and eight yeomanry regiments.

The aftermath of more than four years of bitter fighting hardly provided a suitable climate in which to recruit a voluntary force — neither the potential recruit nor his employer were fired with any great enthusiasm for another war. With no prospect of hostilities the armed services could hardly expect to be given much priority in the allocation of resources. The Territorial Army was almost at the bottom of the list.

During 1921 the country was plagued by continuous strikes by railwaymen and miners, which aggravated the economic situation. In order to maintain essential services, without the elaborate procedure of a Royal Proclamation followed by Act of Parliament, the Government introduced an emergency measure, creating a Defence Force of volunteers, willing to carry out essential duties. Although this force, during the four months of its existence, fulfilled its purpose, it cut right across the organisation of the Territorial Army and hampered its efforts at rebuilding its strength.

Not long after, the 'Geddes Axe', which had such a drastic effect on the Regular Army, had hardly less effect on the Territorial Army. Royal Army Service Corps units, like their Regular counterparts, were reduced to mere cadres. In 1922 the establishment of the Divisional Train was cut from 288 all ranks to 11 officers and 93 other ranks. This cadre consisted of a headquarters (three other ranks), a horse transport company (37 other ranks), a mechanical transport company (23 other ranks) and supply details (30 other ranks). This was the outcome of a committee on national expenditure under the chairmanship of the Minister of Transport, Sir Eric Geddes, the Director-General of Transportation during the war who had done so much towards the efficiency of the Army's Transportation Services.

With such reduced establishments there was no problem in recruiting. But there were acute difficulties in meeting the administrative commitments of supply and transport, particularly during the annual camp, with such few numbers. Realistic practical training was almost an impossibility within the limited resources available.

Nevertheless progress was made in building the morale of the resurrected Territorial Army, by concentrating on theoretical training and the sporting and social activities of units, when the General Strike threw another spanner in the works. All training was cancelled for that year and further reductions were made in funds available. The prevailing optimism in the League of Nations and the Locarno Pact provided a ready excuse for cutting the Defence budget. In the face of rising costs of mechanisation for the Regular Army, demands for more expenditure on the Territorial Army received a cold reception.

It is surprising that, under these conditions, any faith in the future of the Territorial Army remained. It seemed that the cry of 'jam to-morrow, but never jam today' was only too true. There were, however, a few prominent personalities who did a great deal to keep the flame alive. The memory of what the Territorial Force had achieved in the war helped buoy up their hopes. Three members of the Maude family served with great distinction in the Royal Army Service Corps (TA). Colonel A. H. Maude CMG DSO TD DL joined the Army Service Corps in 1908 on the formation of the Territorial Force, served through the war as senior supply officer of the 47th Divisional Train and became its commanding officer after the war. As an eminent journalist he wielded considerable influence. He was a great champion of the Territorial Army, gave unfailing support to the Royal Army Service Corps and became a vice-patron of the Regimental Association. His brother and son also had distinguished careers in the Army Service Corps. The latter, Angus, became Conservative MP for Stratford-upon-

Avon and a junior Minister in Mr Macmillan's administration. He was knighted in 1981 and created a life peer in 1983.

Sir Arthur Holbrook KBE VD JP DL was another staunch supporter of the Territorial Force from its inception in 1908. He joined the Army Service Corps at the outbreak of the Great War. After retiring from active service on his seventieth birthday, he became Conservative MP for Basingstoke from 1920 to 1929, during which critical period he supported the re-organised Territorial Army. He became Honorary Colonel of the 44th (Home Counties) Divisional Royal Army Service Corps (TA) in 1929 and held the appointment until May 1936. He was president of the United Royal Army Service Corps Dinner Club, formed to maintain links forged between Regular and Temporary officers during the war. Two of his sons, Colonel Sir Claude Vivian Holbrook CBE and Colonel Arthur Ernest Holbrook DSO, who was Deputy Director of Supplies at GHQ BEF in France in 1939-40, served with dinstinction in the Regular Royal Army Service Corps. At a moving ceremony at Feltham in August 1932, three generations of Holbrooks were on the saluting base: Colonel Sir Arthur Holbrook, Honorary Colonel the 44th (Home Counties) Divisional Train RASC (TA), Lieutenant Colonel A. E. Holbrook, Commanding Officer Royal Army Service Corps, Feltham, and Second Lieutenant Derek Holbrook RE, who acted as ADC to his grandfather.

Between 1924 and 1930 the Regular Royal Army Service Corps was in transition between horsed and mechanical transport. The horse transport units were being gradually replaced by mechanical transport. This was a time of reappraisal, when the Corps was trying to adjust itself to the new concept of mechanised warfare. The Territorial Royal Army Service Corps units were not directly involved in these developments. But they watched from the sidelines with the greatest interest as the outcome would significantly affect their future.

Apart from mechanisation other important changes were taking place. Royal Army Service Corps training was rationalised and more realism introduced. Closer links were established between Royal Army Service Corps companies and the divisions which they would support on mobilisation. A critical review was undertaken of the organisation for war of the Corps, as a result of experience in recent large-scale Army manoeuvres.

All this created an air of uncertainty in the Territorial Army. Although it was plain to all that the horse was about to give way to the lorry, Territorial Army units were expected to train with horse transport. The 52nd (Lowland) Divisional Train received no mechanical transport at all until late in 1923, when two motor lorries were issued.

By 1929 the Regular Royal Army Service Corps was completely mechanised, the title 'Divisional Train' discarded, and the Divisional RASC reorganised into a headquarters, an ammunition company, a supply company and a baggage company.

As the Regular Army struggled to meet its commitments with reduced manpower, the importance of the Territorial Army began to be appreciated. 1930 was declared a Territorial Year. Assistance was provided by Regular units for Territorial Army training. Courses for Territorial officers and NCOs were started at the Corps Training College at Aldershot. Territorial officers were invited to become members of the RASC Officers' Club.

The economic crisis in 1931 put another severe brake on expenditure. All Territorial Army training for the following year was cancelled. More reductions were made in Army establishments, although this time the Royal Army Service Corps escaped, owing to the recent disbandment of horse transport units. There was no interference in the mechanisation programme.

It was three years before the Territorial RASC officially adopted the new organisation of the Regular RASC. In 1932 the Divisional Train became the Divisional Royal Army Service Corps (Territorial Army). Each Divisional RASC was organised into a headquarters and three companies, the divisional ammunition company, the divisional baggage company and the supply company, which included supply details and a detachment for the field ambulance. The Territorial RASC Companies were renumbered from 501 to 545 (three for each division) and from 546 to 557 for non-divisional companies, introduced in 1936. The peace establishment in 1933 was still only 11 officers and 99 other ranks, but three lorries, one motor ambulance and three motor-cycles were included, and authority was given for the hiring of four cars and three additional lorries for training in camp. On mobilisation this establishment would need to expand to 38 officers and 1,245 other ranks, with more than 500 vehicles.

The 43rd (Wessex) Divisional Train anticipated this re-organisation by two years. In 1930 they dispensed with horses and organised the unit in three skeleton companies, one each for ammunition, baggage and supplies. Their Honorary Colonel (General G. F. Davies, a former Director of Supplies and Transport) may have had something to do with it. He was succeeded as Honorary Colonel in November 1933 by Major L. Hore-Belisha, MP for Devonport, Financial Secretary to the Treasury, later to become Secretary of State for War.

The Territorial Royal Army Service Corps maintained its close links with the motor industry. Sir Herbert Austin KBE was appointed Honorary Colonel of the 46th (North Midlands) Divisional Royal Army Service Corps (TA) in 1929 and Colonel J. S. Mathew CMG DSO TD of the Dunlop Rubber Company was appointed Honorary Colonel of the 52nd (Lowland) Divisional Royal Army Service Corps (TA).

Territorial Army training was resumed in 1933. The organisation was now the same as that of the Regular Royal Army Service Corps, and was to continue so up to the outbreak of the Second World War. Thoughts were already turning to a further reorganisation, including a separate divisional petrol company and a double echelon Supply Column. This was the organisation eventually adopted in 1936 and included in the new edition of *RASC Training Vol 1* (1937).

After the decision to rearm in 1934 it was soon apparent that one of the main gaps was the lack of any organised defence against air attack. And so in 1935 the Air Defence of Great Britain was brought into being as part of the Territorial Army. Initially it consisted of two anti-aircraft divisions, the first formed by converting Territorial infantry battalions of London and the suburbs into Royal Artillery regiments, equipped with anti-aircraft guns, and Royal Engineer units equipped with searchlights. The battalions came from the 56th (1st London) and the 47th (2nd London) Divisions. In January 1936 the 56th Divisional RASC merged with the 47th Divisional RASC to form the 1st Anti-Aircraft Divisional Royal Army Service Corps (TA). The remaining elements re-formed as the London Divisional Royal Army Service Corps (TA).

Shortly afterwards the 46th (North Midlands) Divisional Royal Army Service Corps and the non-divisional companies 546 and 551 were converted into the 2nd anti-aircraft Divisional Royal Army Service Corps, in support of the 2nd anti-aircraft (Northern) Division, formed from the 46th (North Midlands) Division.

It was soon clear that the cadre strength of RASC units was inadequate for their role in support of the Anti-Aircraft formations of ADGB. Tailor-made establishments had to be produced for the Anti-Aircraft Divisional RASC units. These involved large increases and the need for intensive recruiting drives.

One result was a review of the establishments of other Territorial Divisional RASC units, and a new peace establishment of 280 all ranks and a war establishment of 1,350 all ranks was introduced.

Early in 1938 two more anti-aircraft divisions were added. In September and October, during the Munich crisis, anti-aircraft defences were manned and the anti-aircraft Divisional RASCs formed part of the 50,000 men of the Territorial Army involved in the exercise. This afforded an excellent opportunity of assessing the efficiency of the Air Defence organisation.

Mr Chamberlain's appeal for recruits for the Territorial Army, with the promise that volunteers would escape conscription, if it came, had an immediate response, and drill halls were flooded with applicants. Never have the permanent staff been so busy. In the seven months following Munich the Territorial RASC recruited over 20,000 officers and men. When their establishment of other ranks was increased from 246 to 1,300, the 49th (West Riding) Divisional Royal Army Service Corps (TA) recruited 1,100 men in seventeen days. The Territorial Army was no longer in the doldrums.

The close relationship between Regular and Territorial members of the Royal Army Service Corps was well exemplified by the special visit of the Territorial RASC to Aldershot in August 1938. Honorary Colonels and commanding officers attended a guest night in the Buller Headquarters Officers' Mess, at which the Rt Hon L. Hore-Belisha, Secretary of State for War, was present. This was only the third occasion on which such a function had taken place. The first time was in 1909, when HRH The Duke of Connaught was present, and the second in June 1935.

By the spring of 1939 five anti-aircraft divisions had been formed and two more were in the process of forming. Once more it was decided to man part of the anti-aircraft defences, including the whole of the Greater London area. Although the Military Training Act was passed on 9 May, general mobilisation had not been ordered, and so the plan, which was phased over the months of June, July and August, had to be on a voluntary basis. Such was the climate at the time that there was no shortage of volunteers and the Territorial Army demonstrated once more its readiness to come forward at a time of national crisis.

Another change in organisation took place in March 1939, when the 2nd Cavalry Divisional Royal Army Service Corps (TA) became the Mobile Divisional Royal Army Service Corps (TA). There were now in the Territorial Army one Mobile Divisional RASC, twelve Infantry Divisional RASCs and five Anti-Aircraft Divisional RASCs.

At Corps week in May 1939, Major General Macleod announced that the strength of the Territorial Army was being more than doubled.

As war drew nearer many of the Territorial Army units were in the middle of their training camps. Never had there been such a high attendance. Many of the Divisional Royal Army Service Corps units were up to their full strength of 1,350 and few had a strength of less than 800. Some were still in camp when mobilisation was ordered, and moved straight to their mobilisation centres.

Thanks to the vehicle impressment scheme and the preparatory work done by the RASC Inspectorate of Supplementary Transport, units were able to obtain a large proportion of their vehicles with a minimum of delay and, although the process of mobilisation is never without set-backs, this was one occasion when the transition from peace to war went comparatively smoothly.

In a very short time the Territorial Army was ready to take its place once more beside the Regular Army.

The Supplementary Reserve was first introduced in 1924, to meet a transportation requirement for trained railwaymen, who could perform their civilian trade in the Army with a minimum of military training.

The success of this experiment led to the adoption of a similar reserve for the Royal Army Service Corps, who had a serious problem in filling vacancies in the war establishments of regular units on mobilisation. These units, which had been reduced to mere cadre strength, needed to expand four-fold to reach their war establishment. The existing reserves were quite inadequate. The intention behind the creation of the Supplementary Reserve was to earmark individuals in civilian life who had the professional skills required, and would need little more than military training on call-up. For the most part they were not formed into units (the exception was anti-aircraft brigade companies), but were earmarked for mobilisation posting to particular regular units.

For the rank and file there was no training obligation. The initial term of engagement for a Driver was for two, three or four years. Facilities were available for warrant officers to attend courses at the Royal Army Service Corps Training Centre. Officers attended an initial training course of one month at Aldershot. This was followed by an annual fifteen day training period, spent in continuation training with the regular unit to which the officer was paper-posted.

During the period of rapid expansion after the Munich crisis in 1938, special commissions were offered to Supplementary Reserve officers, to help meet the prevailing shortage of officers.

Although, at the start, there was some scepticism about the value of this reserve, in practice it proved to be of immense value, in helping to solve the mobilisation manpower problem. By June 1939 the Royal Army Service Corps (Supplementary Reserve) numbered 300 officers and 11,000 men.

The Supplementary Reserve differed from the Territorial Army, mainly in the absence of formed units and regular training. This meant that the community life and the social activities of the Territorial Army, which played such a large part in maintaining *esprit de corps*, were missing. Morale could not be based on comradeship and unit loyalty. Men of the Supplementary Reserve owed their loyalty to the regular unit to which they were linked by their mobilisation commitment.

51st (Highland) Divisional RASC convoy halt during 1938 camp; note civilian hired Austin 6 of convoy commander.

OPPOSITE ABOVE: Colonel Sir Arthur Holbrook KBE VD JP DL, Honorary Colonel, 44th (Home Counties) Divisional RASC (TA) with a group of his regimental officers, leaving a levée at Buckingham Palace on 18 March 1936. BELOW: Lt Colonel F. W. Grove RASC leads 46 (North Midlands) Divisional RASC TA through the streets of West Bromwich, on King George VI's coronation. ABOVE: 46 (North Midlands) RASC TA off to camp in 1938. BELOW: Visit of Field-Marshal HRH the Duke of Connaught KG to Militia Training Battalion RASC, Boyce Barracks, Crookham September 1939 — back row: Maj G. Bright Holmes; RSM C. A. Bloxham; Capt H. H. Bruton; RSM M. A. Nutt; Lt T. Marrison; front: Lt Col H. R. Dobb; Brig C. le B. Goldney; The Colonel in Chief; Col R. M. Airey, Equerry to HRH, Maj E. Doolan.

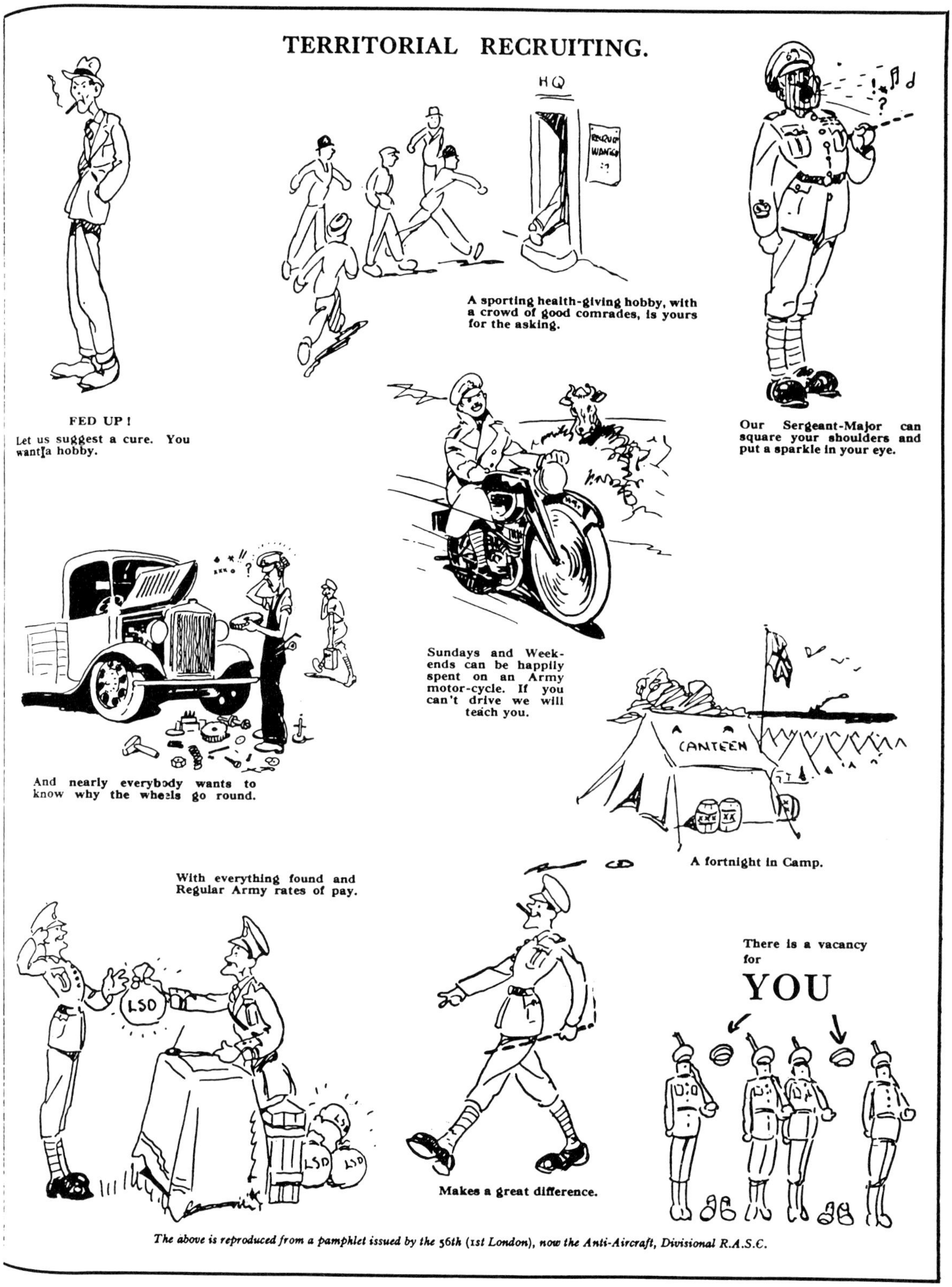

Territorial Recruiting, 1938.

RASC TRAINING *XVIII*

In early days the training of officers and men of the Corps was centred at Woolwich, which claimed to be the senior station of the Corps. After the ASC School of Instruction had been formed at Aldershot in June 1884, the centre of gravity gradually shifted from Woolwich to Aldershot. Woolwich continued to be the home of horse transport and the centre for instruction in all aspects of its operation until April 1929, when the last horse transport unit left Woolwich and moved to Aldershot. Thereafter Royal Army Service Corps training was concentrated at the RASC Training Centre.

The original ASC School of Instruction became the ASC Training Establishment in 1904. The appointment of Commandant was authorised in March 1908 and Colonel S. S. Long became the first holder. In 1920 it was the RASC Training Establishment. In 1924 this was changed to the RASC Training College and in 1931, when a major reorganisation took place, it became the RASC Training Centre.

In its new form the Training Centre consisted of a training battalion, which absorbed the former 'A', 'Q' and 'Y' Depôt Companies, an administrative school, a MT school, with functions previously carried out by 'P' Company, and the Army School of Cookery.

The Officer commanding the training battalion was responsible to the Commandant of the Training Centre for all depôt training and administration. The battalion consisted of: a training company (increased to two in January 1936), which was responsible for the regimental training of recruits, boys, young officers and permanent staff instructors of the Territorial Army, and the higher training of NCOs for promotion and senior appointments; a depôt company for the administration of all the officers and permanent staff; and a horse transport training company (Transport (Animal) Company) for the training in equitation, pack transport and horsemastership of young officers and recruits for the horse transport branch.

The object of forming a training battalion was to amalgamate the previously semi-independent training units and to create a regimental atmosphere which would foster *esprit de corps* in the newly-joined officer and recruit.

The administrative school was responsible for training officers (Regular, Supplementary Reserve and Territorial Army) and permanent staff instructors for the TA in RASC subjects, for higher administrative training and the specialist training of clerks. The MT school, which absorbed the old 'P' Company, was responsible for the training in mechanical transport of officers, warrant officers and NCOs of the RASC and other Arms.

The Army School of Cookery which was commanded by an RASC officer, with a civilian chief instructor, implemented the policy laid down by the Quartermaster General.

Recruit Training

Recruits were provided for the Corps through the normal recruiting organisation of the Army, according to the quotas which were currently in force. After they had been attested at the place of enlistment, recruits were sent to Aldershot, where the training battalion gave them educational, medical and dental examination. Those who had been enlisted as tradesmen were given a trade test. After final approval the recruit was accepted into the Corps and started his twelve weeks regimental training in one of the training companies. Recruits were formed into squads and remained together throughout their regimental training. Each squad bore the name of a distinguished Corps officer, with the object of inculcating a spirit of competition and pride in the Corps. (The original nine names were: Boyce, Carter, Clayton, Connaught, Ford, Heath, Hobbs, Landon and Ward. Six more were added later: Bridge, Bunbury, Crofton Atkins, Gibb, Grattan and Tarver. Finally, as The Training Centre expanded, the names of Paul and Wellington and the four VCs, Dalton, Masters, Morley and Murphy, brought the total to 21.) Military training consisted of drill, physical training, weapon training, educational training and interior economy. Squad NCOs were specially selected and remained with their squads for the whole period of training. At the end of twelve weeks a passing-out parade was held. By this time the recruit had reached a high standard of discipline and physical fitness and had a good grounding in skill-at-arms.

There followed a period of 'specialist' training. Clerks and technical MT clerks received their training under the chief instructor of the administrative school and the MT chief instructor. Butchers, bakers and issuers were trained in the Aldershot supply depôt. MT drivers and artificers went to Feltham, the former to the RASC driving school (until 1936 'R' (Depôt) Company) for a course of eight weeks, and the artificers to the heavy repair shop.

In March 1939, in order to compete with the greatly increased intake of recruits, the period of regimental training was reduced from twelve to nine weeks.

Training of Apprentice Tradesmen

The training of enlisted boys as artificers was originally the role of 'P' Company at Aldershot, which was formed as an artificer training company. The importance of this company was reduced when the technical training of boys was taken on by the Army Technical School (Boys) at Chepstow. This school was first formed on 25 September 1923. Until its accommodation at Chepstow had been completed, the school was located in Buller Barracks, Aldershot. Major F. P. Barnes DSO OBE RASC was the chief instructor until 1926. More than a third of the instructional staff was provided by the Royal Army Service Corps. The school provided artificer training for all the technical arms. A little less than ten per cent of all the boy apprentices trained at Chepstow joined the RASC. This was a valuable source of high-grade NCO material, and men who had completed their apprenticeship at Chepstow played an important part in every field of Corps activity.

At the end of 1937, as the Corps continued to expand, the output from Chepstow was insufficient to meet all its needs. It was decided to resume the training of boy artificers at Aldershot. Arrangements were made for 50 boys to start training as fitters at the RASC Training Centre in January 1938. A boys' wing (fitters) was formed as part of No 1 Training Company. The syllabus was based on that of the ATS Chepstow, the boys doing a three year apprenticeship.

This was a prelude to the decision taken in the summer of 1938, that a Boys Technical School RASC would be established in St Peter's Barracks, Jersey, with effect from 1 October. It was agreed that the school would take boys of the RAOC until their own apprentices' school was established at Arborfield.

Lieutenant Colonel W. N. Craig-McFeely was appointed Commandant of the newly-formed Army Technical School (Boys), Jersey, with Captain C. V. Ferrey as chief instructor and Captain A. P. Campbell as adjutant and quartermaster. Three of the successful 1937 football team were among the NCOs appointed — Sergeants Dallas, Stewart and Waite.

The school opened on 6 October with 243 boys, some 100 of whom were RAOC. It was organised on imaginative lines, with six barrack blocks forming 'houses', rather like those of a public school, and appeared to have a most promising future. But fate was to decide otherwise.

After their summer leave in 1939 the RAOC boys left for their own school at Arborfield. Less than a year later, after the fall of Dunkirk, the whole school was evacuated to Blackdown and became part of the RASC Training Centre.

Junior Officer Training

When the war ended there was a considerable number of young Corps officers who had never completed a full-length junior officers' course. These were officers who had been commissioned during the war and subsequently received Regular commissions, or those who had been seconded to other Arms and who had only recently returned to the fold. The first few courses in 1919 and 1920 were entirely taken up with this commitment. In the autumn of 1921 the junior officers' course included the first new intake from the Royal Military College and a University candidate. At that time the course lasted nine months, the first two of which were spent at Woolwich, where the young officer was given instruction in equitation, pack transport and horsemastership, followed by six months at Aldershot, where instruction was given in all other aspects of RASC training. Regimental training was spread over the whole course. Between 1919 and the beginning of 1930 officers were commissioned into the Corps from several sources: direct entry from the Royal Military College, Sandhurst, transfer from other Arms and Services and the Territorial Army, and university candidates who had obtained degrees. There was no direct commissioning from the ranks. Selected NCOs from all Arms and Services were awarded 'Y' cadetships at the RMC and, after completing their training at the College, were commissioned in the normal way into the regiment or corps of their choice, subject to the approval of the regimental colonel.

Conditions for the transfer of officers to the RASC were laid down in 1925 as follows: the candidate must have had not less than two years commissioned service, and would be on probation for one year, which could be terminated earlier in the interests of the Service; transfer was in the rank of lieutenant, with seniority from the date of appointment to the RASC on probation.

Although officers of other Arms were not encouraged by their commanding officers to transfer, unless they had shown themselves to be misfits in their regiment or if they had disobeyed orders by getting married under the age of 30, there were a number of inducements. It was easier to live on a small income in the RASC than in a Cavalry or Infantry regiment. Corps pay, not substantial for a junior officer, was a help. Officers were permitted to marry at the age of 26 (the age at which they qualified for marriage allowance) or earlier, on sufferance. Chances of promotion were much more favourable than in the Infantry. In some regiments there were subalterns with as much as nineteen years service. Slow promotion made the life of an Infantry officer most frustrating. A young officer could see nothing ahead except the same annual training cycle — a future of boredom, only to be relieved by foreign service or some sudden emergency. If he had an interest in motor cars and things mechanical, the RASC had a distinct appeal. Finally, service in the RASC offered a great variety of experience, with a large number of stations both at home and abroad, offering a higher level of responsibility than a young officer in an Infantry regiment could expect.

The RASC benefited from the wide spectrum of experience which transfers brought them. Not only were there representatives of the Royal Artillery, Cavalry and Infantry regiments, but officers of the Royal Navy, Royal Marines and Royal Air Force came to the Corps on transfer. Not all of them survived the year's probation. Some went back to their regiments. Others, like the notorious Lieutenant Baillie-Stewart ('the officer in the Tower') fell by the wayside. But most of them were accepted and made a valuable contribution to the life of the Corps.

In 1926 a scheme was introduced for the attachment of officers of other arms to the RASC for a period of five years. Applicants for such attachments were required to have not less than two years, and not more than five years service at the time of application.

Not many university candidates were commissioned into the Corps during the 1920s but, when the rules for commissions in the Corps were changed in February 1930, the need for more officers from this source became apparent. It was decided that, from 1 February 1930, direct entry of gentlemen cadets from the RMC would cease, and that only subalterns with not less than two years commissioned service would be eligible to transfer, retaining seniority up to five years. This meant that, if a cadet at Sandhurst wished for a commission in the RASC, he had to select some other regiment in which to serve his qualifying service, before transferring to the Corps. Special efforts were made to interest the universities in careers in the RASC. Within two or three years these efforts bore fruit, and by 1934 No 40 JOC consisted of ten university candidates and six transfers, and this kind of pattern was maintained up to the outbreak of war in 1939.

One effect of these new rules was that a larger number of young officers entering the Corps had had no general military training. To overcome this, it was decided that all university candidates would spend a year attached to an Infantry regiment, after the completion of their junior officers' course. Later, this was reduced to six months. At the end of 1938 it was decided that University candidates would do one year's regimental duty, immediately after the end of the JOC, before their attachment to an Infantry regiment.

At the time of the reorganisation of the Training Centre in 1931, newly joined officers came under two categories: (1) those joining direct from universities on first commission; and (2) those transferring on probation from other Arms and Services.

University candidates, having had little previous military training, received a month's preliminary regimental and military training before the start of their junior officers' course. On the course, both university candidates and officers on transfer were given regimental training concurrently with RASC training under the chief instructor. In addition to lectures on RASC subjects, equitation, drill, weapon training, physical training and anti-gas training were included. The course ended with four months instruction at the MT school.

Officers for the Corps continued to come from these two sources (transfers and university candidates) until, in 1939, the urgent need for a greater intake made it necessary to revise conditions. Commissions were once more thrown open to cadets from the RMC, regular commissions were offered to Territorial Army and Supplementary Reserve officers, and a form of temporary commission was introduced. The number of junior officers' courses was increased from two to four a year, and the length of the courses was reduced.

Immediately after the 1914-18 War, the policy was for all officers entering the Corps to receive training in mechanical engineering, up to the standard which would enable them to qualify as AMIMechE. This would entitle them to draw Corps pay at the higher rate. This involved three years of technical training, one year's MT long course at Aldershot, followed by a two-year practical works course at a civilian engineering firm, or one year in a civilian works firm, followed by one year in the heavy repair shops at Feltham.

During the 1920s the dialogue on the future role of the Corps — whether it should become an entirely technical Corps or whether the main emphasis should be on a regimental Corps with mechanical engineering as only one of its responsibilities — went back and forth for ten years. Eventually the regimental school won the day. It was decided that only those junior officers who already had engineering degrees or showed a particular aptitude for mechanical engineering should receive the full three years training. The remaining officers would receive a short MT course, which would be sufficient to qualify them for the efficient operation of mechanical transport. In each company there would be a technically qualified workshops officer. No 31 JOC was the last course to attend a long MT course, *en bloc*, in April 1931.

In addition to a number of transfers from Commonwealth Armies, particularly from the Australian Army, officers from Commonwealth and Allied Armies were admitted to junior officers' courses on attachment. The Citizen Forces of South Africa, the Egyptian Army, the Indian Army Service Corps, the Iraq Army and the Royal Canadian Army Service Corps (Lieutenant G. E. R. Smith, who later became DST and Honorary Colonel Commandant of the Royal Canadian Army Service Corps) were all represented during this period.

Having satisfactorily completed his junior officers' course at the RASC Training Centre, an officer became entitled to Corps pay, which in 1934 amounted to the 'princely' sum of 1s 4d a day for a subaltern, with an increase each time he was promoted to a higher rank. His further training now became the responsibility of his commanding officer and his company commander. Under their direction he might be sent on a course, if he could be spared, at one of the Army schools of instruction or at the RASC Training Centre. But for the most part, his training would consist of practical training in his unit, or participation in schemes devised by his commanding officer or formation training organised by the district or divisional commander.

Continuation Training

Before the reorganisation of the RASC Training Centre in 1931, little 'career' training was provided for officers and NCOs of the Corps. All this was changed, when the responsibility for the training of the RASC was transferred from the Quartermaster General to the Director General of Military Training. Colonel D. C. Cameron, as commandant, appreciated the need for constant study and discussion of the problems which were likely to face the Corps in a future war. With this end in view he inaugurated, in 1929, the first of a series of war courses, which were designed to study, in the most practical way, the principles on which a Field Force is organised and dispatched overseas, with particular reference to the part played by the RASC. To give greater value to the discussions, attendance on these war courses was open to officers of all arms of the Service and the Royal Air Force, in addition to senior captains and majors of the Royal Army Service Corps.

Close co-operation with the Staff College was maintained, both in the preparation of these courses and in the conduct of the courses themselves. The general staff officer appointed to the Training Centre looked after the operational aspects of the courses, and ensured that administrative instruction conformed to the general training policy of the DGMT.

At the time, these courses provided the only opportunity, outside the Staff College, of studying the administrative and logistic problems of war in a realistic manner. The courses were of two kinds. The winter war course was designed to study the problems of the following phases of war:

(a) mobilisation and embarkation;
(b) the preparation of a base and the landing of a field force;
(c) the maintenance of a force on a long L of C.

The summer war courses, which had greater opportunities of outdoor studies, concentrated on the RASC problems connected with the carriage and distribution of ammunition, supplies and petrol from railhead to delivery points, and the general organisation of RASC field units.

These war courses not only aroused a great interest among those selected to attend, and provided a new stimulus to their professional studies, but also gave a keen incentive to the staff of the Training Centre to keep their administrative doctrine under constant review.

An advanced supply course was introduced at the same time as the war courses, which covered much the same ground as the courses held at the London School of Economics for Army Officers, which a number of RASC officers had attended. The course covered such subjects as the manufacture and inspection of foodstuffs, economic and accountancy problems, commercial law and factory management, and included visits to a number of civilian establishments.

It was laid down in King's Regulations that 'the RASC Training Centre is established primarily for the purpose of the military training of officers and other ranks of the RASC (Regular Army, Supplementary Reserve and Territorial Army), including supply and transport duties in peace and war.'

In order to comply with this charter, the Training Centre organised some fifty courses a year, covering all aspects of RASC training, about half of them on general and supply subjects, and the other half at the MT school of instruction.

In addition to courses designed to give practical instruction in all aspects of RASC duties to officers and NCOs of the Corps, courses were arranged for members of other Arms, the Royal Air Force (for whom a special supply course was arranged) and officers of Commonwealth Forces.

The RASC Training Centre had become the nerve-centre of the Corps and as such was of the greatest importance in the study and development of administrative systems for the efficient maintenance of the Army in war. It was largely responsible for the reorganisation of the RASC field units, which was published in *RASC Training, Vol I* in 1937, and which was virtually unchanged when the Army went to war in 1939. The importance of the centre was reflected in the quality of the officers selected to fill the appointments of commandant and chief instructor of the Training Centre. Mention has already been made of Colonel Cameron. Other names which spring to mind are: Colonels W. N. White and M. S. Brander (commandants), Lieutenant Colonels H. N. Foster, G. E. Badcock, H. P. Raymond, P. L. Spafford and H. R. Kerr (chief instructors) and Lieutenant Colonels M. S. Brander and R. T. Snowden-Smith and Majors F. S. Clover and H. C. Goodfellow (MT chief instructors). Three of these later became Inspectors and five became Directors of Supplies and Transport. Snowden-Smith, as a young subaltern, had been one of the Army's most distinguished amateur pilots and in 1911 came second in the Brooklands to Brighton aeroplane race in a Farman biplane.

The RASC Training Centre played a leading part in many of the Aldershot Command functions. For three consecutive years Lieutenant Colonel Spafford was the chief steward of the Aldershot Horse Show, and made use of a number of Training Centre officers as ring stewards. Two of the recruit squads each year had their training extended by a month, to enable them to take part in the Aldershot Tattoo. They usually represented the enemy in the main event, such as the French Army in the pageant of 'The Crossing of the Douro' from the Peninsular War, and also played a part in three or four of the other events, including the grand finale. Rushmoor Arena became the stage for the tattoo in 1923, with a capacity of 84,000 (later increased to 85,000). In 1933 nearly half a million people came to see this great spectacle.

Higher Training

Collective training was not resumed after the war until 1924. During the interval great advances had been made in mechanisation. Large-scale exercises provided a good opportunity of testing the capabilities and efficiency of mechanical transport and paved the way for the reorganisation of the Divisional RASC on an entirely mechanical basis.

Owing to the heavy administrative commitments involved in the preparation of large-scale manoeuvres, most of which fell on the RASC, and the comparative shortness of the exercises (seldom more than about four days) it was difficult to make a realistic assessment of the administrative problems of maintaining large mobile forces, but many valuable lessons were learnt, and it became clearer what sort of organisation would be required for the satisfactory maintenance of a mechanised army. These manoeuvres continued annually until 1929, allowing modifications in organisation to be tried out, but the depression of the early 1930s caused the cancellation of large-scale formation training above divisional level until 1935, when Aldershot Command and Southern Command combined together in the largest manoeuvres which had yet taken place.

The organisation of the RASC field units which evolved from these training exercises, and the studies which had been taking place at the RASC Training Centre, proved their worth when war broke out, and few major modifications were found necessary, except to meet the abnormal conditions in some overseas theatres. In the Middle East, for example, standard systems had to be adapted to meet desert conditions. The problems of driving in soft sand, of desert navigation, of the lack of any shelter or natural protection from observation, all demanded special study and the adaptation of conventional procedures to the particular conditions. The use of sun-compasses, sand channels and mats, and camouflage nets for concealment formed an essential part of the training for desert warfare. The carriage of water was as important as the carriage of supplies.

During the 1929 manoeuvres in England, an improvised water tank company RASC was formed for the provision of water for the 2nd and 3rd Divisions, training between Warminster and Bulford. This was no imaginative idea of the exercise planning staff but was due to a prolonged drought. A number of new Morris 6-wheeled chassis were specially fitted with water tanks and pumps at Feltham.

During its short life the company made a dump of 90,000 gallons at Imber and delivered about 100,000 gallons of water between 8 and 15 September.

In Egypt the provision of water was no new problem. The standard container was the 12½ gallon camel tank. Twenty-five of these 'fanaties' constituted a load for a Thornycroft six-wheeler and was the equivalent of two 150-gallon tanks on a Morris six-wheeled vehicle. A special feature of the Cairo Brigade 1931 manoeuvres was the carriage and delivery of water in mechanical transport. Previously camel transport had been used.

Later in the same year a special MT trial was carried out in the desert to Bir Hooker, in which Captains C. F. Neve and L. A. Tudor took part. However, the authorities were not entirely satisfied that reliance could be safely put on MT for all maintenance in the desert. Command manoeuvres in 1933 included the formation of a camel transport unit for the supply of a force of 4,500 men and 800 animals for three days. The unit consisted of 1,000 camels and 500 camel drivers.

Further trials of MT in the desert were conducted from Khartoum in 1933 — a War Office experimental convoy — and from Cairo in the same year, across the desert to Siwa with armoured cars. In 1938 a comprehensive trial took place in conjunction with Dunlop, mainly to test new models of desert tyres. In the same year the first supply by air exercise was held in the desert.

The emergency following the Italian invasion of Abyssinia gave an impetus to training in the desert, and the comparatively small force stationed in Egypt (limited to 10,000 under the treaty of 1936) reached a high state of training for desert warfare by 1939.

Co-ordination of Training with Commonwealth Forces

In order to ensure a close liaison with the equivalent allied Army Service Corps of Commonwealth countries, a system of exchanges was arranged with the Royal Australian Army Service Corps and the Royal Canadian Army Service Corps. These exchanges were normally for two years and the officers involved carried out exactly the same duties as their counterparts would have performed. The exchanges with the RCASC were particularly successful. In addition to Lieutenant G. E. R. Smith, Major J. E. H. Tidswell RCASC and Lieutenant J. L. Sparling RCASC came to England on exchange, and Captains N. G. Duckett, R. H. Elliot and C. H. Collinssplatt went to Canada. Captain N. B. Macdonald was attached to the RASC in exchange for Captain G. E. C. Rosall, when the Abyssinian crisis occurred and he spent part of his attachment in the Western Desert. He later became DST of the RCASC.

The Staff College

RASC officers were not always encouraged to take the Staff College examination, mainly on the grounds that successful candidates would serve outside the Corps, and that the Corps would consequently lose some of its best officers; it was rather left to the individual officer to get there by his own efforts. 'Studying for the Staff College' was often regarded as an excuse for avoiding some tiresome duty. In any case, the number of vacancies reserved for the RASC was minimal, and competition for a place fierce. However, this rather parochial attitude gradually changed, and by the late 1920s the value to the Corps of a number of staff-trained officers was fully appreciated; officers were actively encouraged to take the examination. At the end of 1923 there was only one vacancy reserved for the RASC at the Staff College. Two years later, out of a total of sixty-three vacancies, forty-two were competitive, of which one was reserved for the RASC. The remaining twenty-one were for nominations received from the British Army, the Indian Army, the Royal Navy and the Royal Air Force. Nevertheless, through these years the Corps managed to secure at least two vacancies on most courses. In 1925 Captain P. S. Whitcombe and Captain H. N. Gallacher were students. Four candidates sat for the following course. Captains H. H. Dempsey and H. M. Gale secured places. In 1928 Captain C. S. Vale gained a competitive vacancy and Captains T. W. Richardson, H. R. Kerr, C. E. S. Dobbs and R. T. Cooke all qualified. Dobbs later secured a nomination. Cooke was successful a year later, while Kerr had to wait two more years before he obtained a place. In March 1929 total vacancies reserved for the technical arms was reduced to six — RE four, R Sigs one, RASC one. Captains H. M. Whitty and C. M. Smith qualified in August 1929. In that year there were three RASC students at the college (Vale, Dobbs and Cooke). Thereafter, there were usually at least two RASC officers on Staff College courses. Finally, a reorganisation of the Staff College in 1938 made allowance for three RASC students at the college.

The presence on the staff of senior RASC officers was of great value to the Corps in maintaining close touch with other Arms and Services, and in securing a better understanding among commanders and staff of the realities of administration. In 1933 six important senior staff appointments were held by RASC officers: Brigadier i/c Administration, Egypt (Brigadier J. C. Browne CMG DSO ADC), AA & QMG i/c Administration, Malta (Brigadier E. F. Falkner CMG DSO psc), AQMG, Scottish Command (Colonel G. H. Gill CMG DSO psc), Brigadier i/c Administration, Scottish Command (Brigadier J. L. Weston DSO psc), AA & QMG, 2nd Division, Aldershot (Colonel A. F. G. Pery-Knox-Gore DSO psc); and AQMG, Northern Command (Colonel P. B. Fraser DSO OBE psc). In addition, a number of junior staff appointments were held by RASC officers. One effect of this increased representation on the staff was that, before and during the 1939-45 war, the Royal Army Service Corps provided instructors for both the Staff College at Camberley and the Middle East Staff College at Haifa.

ABOVE: Training in the elementary principles of the four-stroke internal combustion engine. BELOW: Driver-training at R Company Driving School Feltham in 1926; traffic lights had been recently introduced, and the drivers of the Morris and Thornycroft six wheelers are practising as the lights change.

ABOVE: A Platoon, 45 Company RASC, driving Morris Commercial 6 by 4, relaxing before setting off on an exercise in 1933. Dvr A. C. Pearce of 38 Company RASC pays a visit on a Douglas motorcycle. BELOW: The Drill and Weapon Training Staff, Training Battalion RASC Aldershot, 1933; top: LCpl W. Goodwin; Cpl F. Gilks; LCpl S. Shaw; LSgt S. Sidney; LCpl J. Elphick; LCpl R. Roberts; Cpl C. Simpson; Cpl F. Townsend; bottom: Sgt J. Senior; Sgt E. Gayton; RSM F. Marrison; Lt F. J. C. Goodyear-Pain; Sgt J. F. Smith; Sgt E. Bellow; Sgt E. Bostock.

ABOVE: An intake of boy enlistments standing in early October 1938 at the quay at St Helier, Jersey, waiting to be moved to the new technical school established in St Peter's Barracks. BELOW: A group of apprentice tradesmen standing at the quay at St Helier; all were evacuated before the German occupation of the Channel Islands. One of them was Peter Blunt, later Major General P. Blunt CB MBE GM, Transport Officer in Chief (Army), 1973-76.

A convoy of Dennis vehicles with solid wheel tyres.

MECHANICAL TRANSPORT XIX

The most revolutionary change to affect the Royal Army Service Corps during these two decades was the total eclipse of horse transport by mechanical transport. This process started in 1902 and was complete by 1930, when the last service horse transport Company was disbanded.

Experience in the South African War in the use of steam tractors, mainly as gun-towers, had demonstrated that there was a military future for mechanical transport. It was decided in 1902 that further development would be the responsibility of the Army's transport Corps, the Army Service Corps, instead of the Royal Engineers, who had previously been concerned with experiments in this field.

In 1903, Corps order 32/1903 made the historic announcement: 'No 77 (MT) Company will be formed'. This company was raised at Chatham early in 1903 under the command of Captain (Brevet Major) C. E. McNalty and consisted, initially, of one motor car and twelve traction engines, taken over from the Royal Engineers. This was followed later in the year by the formation of 78 (MT) Company at Aldershot under the command of Captain H. R. Hayter. The company numbers followed the sequence of the seventy-six Horse Transport Companies already in existence or authorised. Thereafter Aldershot became the centre of further development of mechanical transport. 52 (Depôt) Company was formed as the original driving school, heavy repair shop and MT stores depôt. In 1904 Major W. E. Donohue transferred from the Army Ordnance Corps to become the first Inspector of Mechanical Transport. From the beginning, MT instruction was given a high priority at Aldershot, and short and long MT courses and works courses with civilian firms were originated as essential elements of the mechanical transport branch of the Corps.

Development of MT was badly handicapped for several years by the conviction at Army Headquarters that petrol was too dangerous a fuel for military use, particularly for the carriage of ammunition. Research and experiment were restricted to the steam engine, and on internal combustion engines designed to run on paraffin. This erroneous belief came from a report by civilian experts that petrol was an unsafe substance, particularly in tropical climates. It was ten years before this crippling ban was officially withdrawn.

In presenting the Army Estimates in March 1911, the Secretary of State for War made a far reaching statement on the substitution of mechanical transport for horse transport, which spelt out in great detail the proposed re-organisation of the Army Service Corps in the field, in order to take full advantage of recent developments in mechanical transport. This decision was largely the result of experience gained from manoeuvres in France and Germany, and at home. The proposals were based on making the maximum use of mechanical transport behind the forward areas, wherever the roads were suitable. Their immediate practical effect would be a reduction of twenty horse transport companies and their replacement by twelve mechanical transport cadre companies, making a saving of 795 personnel, 2,686 horses and

1,908 vehicles in the Army Service Corps, and 4,770 personnel and 5,930 horses in the Royal Artillery. Provision would include the purchase of a proportion of the new mechanical transport vehicles, and funds for subsidising the necessary commercial vehicles.

There was no question of holding in peace time the war establishment of vehicles. So it was necessary to influence design trends in the commercial field, so that sufficient and suitable civilian vehicles would be available on mobilisation. The War Department subsidy scheme was launched in 1911 and Captain H. N. Foster was appointed chief inspector of subsidised transport vehicles. The first scheme was provisional, and embraced both 3-ton and 30-cwt commercial vehicles then in use. The period of the subsidy was three years, and the total payment £110 per vehicle.

The War Office Mechanical Transport Committee, of which Captain Foster was a member, was charged with the examination of the development and use of mechanical transport for military purposes. This committee, in close collaboration with manufacturers, evolved a specification, which was published in 1912, for a War Department subsidy type vehicle. Its technical objects were: (i) to embody the best practices of the makers, (ii) to standardise manipulation and control, and (iii) to minimise as far as possible the variety of spare parts to be carried in the field.

By the summer of 1914 nearly 900 vehicles had been secured under the scheme, which made a considerable contribution to mobilisation requirements and greatly reduced the numbers required and the time occupied in obtaining the deficiency by impressment. The peace establishment of twenty mechanical transport companies in 1914 included 80 motor lorries and tractors, 20 motor cars and vans, and 15 motorcycles. The balance required on mobilisation was 1,061, 193 and 116, making a total of 1,370. The success of the scheme set a pattern followed throughout the inter-war period.

During the 1914-18 war mechanical transport more than fulfilled the expectations of those who worked for its introduction. In France more than 32,000 3-ton petrol lorries were in service. There were 13,000 motor cars, vans and ambulances, 1,000 steam lorries and 1,000 gun tractors. In Mesopotamia large numbers of Fords were used and in East Africa the Autocar was predominant. The totals in all theatres on 11 November 1918 were 56,659 motor lorries and tractors, 23,133 motor cars and motor vans, 7,045 motor ambulances and 34,865 motor cycles. Before the war the Army Service Corps MT inspection branch had been responsible for technical research and development, under the direction of the Mechanical Transport Committee, of which Captain T. M. Hutchinson (later Colonel T. M. Hutchinson DSO OBE MIMechE) became secretary in 1914. He had joined 77 (MT) Company when it was first formed and then moved to the MT depôt at Aldershot, and became MT instructor at the training establishment. Later he was head of the MT inspection branch and appointed chief Inspector of Mechanical Transport in 1918. After commanding the heavy repair shop at Chaklala for four years he became the mechanical transport adviser to the India Office in the rank of Colonel in 1928, when the Royal Army Service Corps ceased to be involved in India. Other officers closely concerned with mechanical transport in the early days were Captains H. N. Foster and H. R. Lever, and Lieutenants M. S. Brander, G. C. G. Blunt, G. H. Gill and E. S. Unwin.

This responsibility continued to be borne by the Royal Army Service Corps MT School of Instruction after 1918, under the expert guidance of Lieutenant Colonel H. Niblett, a civilian professor of mechanical engineering, who was employed as an instructor at the School, after serving as a temporary officer in the Corps. The School was responsible for developing the ideas of the War Office MT Advisory Board, set up at the end of 1920, as a successor to the pre-war MT Committee. This Board was responsible for the development of all wheeled vehicles used by the Army. The chairman of the Board was the Director of Supplies and

Transport, Major General Sir E. E. Carter, and the civilian members were Professor W. E. Dalby FRS, Sir Henry Fowler KBE, Lieutenant Colonel H. Niblett CBE DSO and Mr L. A. Legros OBE. Its terms of reference were 'Not only to act in an advisory capacity in technical problems in connection with mechanical transport for Army requirements, but will secure a close liaison with the motor industry, so that the Army may be in touch with the progress of mechanical transport throughout the world'.

At the end of the war most of the remaining stocks of load carrying vehicles were the 3 to 4-ton lorry on solid rubber tyres. This was the type specified under the pre-war subsidy scheme. It was an economical and reliable performer when road conditions were good, but unsuitable for negotiating soft and rough ground. Its cross country ability was negligible.

The first task of the Royal Army Service Corps experimental workshops at the MT School of the Training Centre was to design a vehicle with much better cross country performance. The light Fiat lorry and the Ford van had shown that vehicles with a lighter axle weight and fitted with pneumatic tyres had a far better performance.

A War Department specification was drawn up for a type of vehicle which would be suitable as a replacement for the horse transport of the Divisional Train. The main features of the specification were: (i) weight as low as possible; (ii) high torque-weight ratio; (iii) high speed; (iv) large size pneumatic tyres; (v) standardisation of various features of design, control and equipment, and (vi) suitability for commercial use.

These were the specifications of what became the War Department subsidy-type 30-cwt lorry. These requirements were only met by reducing the useful load to 30-cwt, by the use of high-grade materials, by most careful design, eliminating all non-essential material, and by the employment of pneumatic tyres. Two units, which were sent to Chanak during the emergency in 1922, were equipped with these vehicles and reported that they were very satisfactory. There was, at first, some reluctance on the part of the manufacturers to produce such vehicles. The solid tyre seemed to many quite adequate and more economical. However, eventually a small number of firms agreed to go into production and a subsidy scheme was launched in 1923. By 1926 the full quota of 1,000 approved vehicles in commercial use had been reached and were ready for immediate use by an expeditionary force, if required. In addition, a large number had been purchased for the peace establishments of Royal Army Service Corps units. They were soon in steady manufacture and a considerable overseas market had been created. It was generally admitted that this War Department specification of 1922, backed by the subsidy and the co-operation of the manufacturers and civilian users, had had a most beneficial effect on subsequent commercial vehicle design. As Captain B. H. Liddell-Hart, military correspondent of the *Daily Telegraph*, wrote after watching a demonstration of new vehicles at Aldershot: 'Commonly throughout history military progress has followed, often tardily, in the wake of civil progress, but in the adoption of pneumatic-tyred transport vehicles the Army has for once led the way'.

Although the Royal Army Service Corps experimental workshops could be well satisfied with the success of the 30-cwt subsidy lorry, it fell short of the ideal in one very important aspect. In order to meet the need for cross-country performance, it had been found necessary to reduce the useful load to 30 cwt. This was no more than the GS wagon could carry and, compared with the 3-ton lorry, meant twice as many vehicles, men and road space for the movement of a given tonnage – all most undesirable factors in war. It was therefore necessary for designers to concentrate on the development of a vehicle with the same performance as the subsidy 30-cwt lorry, but with the following additional characteristics: (i) increased useful load capacity; (ii) lower intensity of pressure on the ground; (iii) high speed; (iv) reasonable first cost; (v) economy in running and maintenance; (vi) ease of handling, and (vii) attractiveness for the commercial user at home and overseas.

At first sight it appeared that a tracked or semi-tracked vehicle might meet the main requirements, but a consideration of (iii), (iv), (v) and (vii) was more than enough to rule that out as a general load carrier. Any tracked vehicles essential to the Army's requirements would have to be held in stock in peacetime.

The solution lay in the provision of a greater area of contact with the ground per unit of load carried. Attention was directed thus towards a means of providing more driving wheels, and it was decided that a rigid frame six-wheeler designed to drive on at least two axles, presented the most promise.

Various foreign designs were studied and a French Renault 6-wheeler, which had had outstanding success in crossing the Sahara, was acquired for testing by the RASC training establishment. Trials showed that the investigations were on the right lines, and experiments were immediately started on the design of a 6-wheeler which would meet the specifications.

No vehicle of this type was in commercial production at the time (early 1925) but the experiments carried out had been so promising that it was decided to approach manufacturers with a view to constructing a prototype. Some firms, who already held the War Department certificate for the subsidy type 30-cwt lorry, were persuaded to take an interest, and two of them (Morris and Thornycroft) each built a vehicle. These were moderately successful and it was decided that with certain modifications and improvements, they would satisfy the requirements.

Later in 1925 provisional specifications were published by the War Department. More firms became interested and orders were placed with four firms during the financial year 1925-26.

The first trials of the 6-wheelers were held in April 1926 and continued throughout the training season at Aldershot. During brigade and divisional exercises in August and September the 'light' model (30-cwt) proved acceptable as a replacement for horse transport in Cavalry and Infantry 'B' echelon, and the 'medium' (3-ton) model for use in Divisional Royal Army Service Corps units.

By the end of the year (December 1926) these vehicles were in full commercial production and numbers were successfully operating overseas in India, South Africa, Egypt, the Sudan and Persia.

An 'overall non-skid chain' was provided, which could be quickly fitted over the driving wheels, converting the vehicle into a semi-tracked vehicle. In the following year Lord Montagu of Beaulieu took part in the trials of the Thornycroft 6-wheeled lorry (Standard A3 type, 2½ tons) in an expedition from Beirut to Baghdad, and expressed himself well satisfied with its performance over desert conditions. Lord Montagu had been adviser on mechanical transport services to the Government of India from 1915 to 1919.

In April 1927, 38 (MT) Company was specially formed at Aldershot for service with the Shanghai Emergency Force, and was equipped with the new Morris 30-cwt 6-wheeler. On the way, the Company, with the rest of 15 Infantry Brigade, spent some time in Hong Kong and Kowloon, where they occupied themselves in training and in putting their sixty Morris 6-wheelers to all sorts of severe tests, including hill climbing, crossing flooded paddy fields and driving through shallow sea water. Lieutenant Colonel Badcock reported that the vehicles surmounted all these obstacles most effectively. As a result of one demonstration, a business firm in Canton immediately ordered ten of the vehicles.

Thus the rigid frame 6-wheeled vehicle was accepted as the standard Army vehicle in its light, medium or heavy form. Three years later it was adopted as the standard vehicle by the Canadian Army Service Corps.

Another development of the experimental workshops, which attracted a great deal of attention when it was demonstrated at Aldershot in the autumn of 1925, was the 54 hp

Thornycroft Hathi 4-wheeled drive tractor and trailer, a most versatile equipment with very good cross country performance, which could be put to a great variety of uses. An interesting experiment was the development of a 3-wheeled motor cycle, with drive on both rear wheels, which was designed to give superior cross country performance. It succeeded in doing what was asked of it, but was never adopted as a working model. Already in 1924 work was in hand on developing special 'balloon' tyres in conjunction with tyre manufacturers, which eventually proved so successful as 'desert' tyres.

The marked co-operation between the War Department, represented by the Royal Army Service Corps experimental workshops, and commercial motor manufacturers and users, was a good augury for the future and ensured that large numbers of the same types of vehicle which the Army used in peace could be rapidly taken into service in the Army in the event of future mobilisation. Such co-operation was strengthened by such appointments as that of Colonel J. S. Mathew CMG DSO TD of the Dunlop Rubber Company as Honorary Colonel of the the 52nd Divisional Train TA, and Sir Herbert Austin KBE as Honorary Colonel of the 46th (North Midlands) Divisional RASC TA. The former contributed a great deal in liaison work with the War Department, and the Government of India, in the development of cross country transport and pneumatic tyres. More than a dozen motor manufacturers accepted young Royal Army Service Corps officers on their two year works courses, which helped to maintain contact between the Corps and the commercial world.

In February 1928 an important reorganisation took place. Until then the RASC Training Establishment had carried out most of the experimental work concerned with military wheeled transport. Now, the rapid expansion of mechanical transport in the Army, and the impending extension of mechanisation to the first line transport of the Cavalry and Infantry, necessitated a change in the organisation which had resulted from the Esher Committee's recommendations in 1904, and which had remained substantially unchanged.

Because the Royal Army Service Corps was the chief user of mechanical vehicles for general transport purposes, the responsibility for certain duties connected with research, design and experiment, as well as the provision and maintenance of a large numer of vehicles of non RASC units, had remained with the Corps. Now these responsibilities would be transferred from the Quartermaster General to the Master General of the Ordnance.

As a result, the Master General of the Ordnance became responsible 'for the research, experiment, design, manufacture, inspection, provision, storage, issue and repair (other than first line repair) relating to ... mechanically propelled vehicles of all kinds except ...for the provision, inspection, storage, issue and repair of all MT vehicles on the establishment of RASC units, including MT vehicles in Medical units', which continued to be the responsibility of the Quartermaster General. The latter also remained responsible for the impressment of vehicles on mobilisation, and for any subsidy schemes in respect of vehicles in peace. This meant that the responsibility for all research and design, both for wheeled and for tracked vehicles, was centralised in the MGO's department.

This reorganisation recognised the growing mechanisation of the Army and the increasing use of mechanical transport by other Arms than the RASC and was an entirely logical development. It was not universally popular within the Corps, as it involved the loss of 10 officers, 146 other ranks and 87 civilians. However, this was partially offset by the creation of some new extra regimental appointments. But it did mean the loss of the experimental workshops, which could claim a most successful record of achievement since the war. Lieutenant Colonel G. le Q. Martel DSO MC RE wrote in *In the Wake of the Tank*: '. . . the development of the 6-wheeled lorry was originally launched by a small band of enthusiasts in the RASC with little outside support or encouragement.'

The functions of the MT Advisory Board were absorbed by the new 'Mechanical Warfare Board' which later became the 'Mechanisation Board'. The Royal Army Service Corps had representatives on the Board and also in the War Office MGO6 branch (the equivalent of QMG3) and the new Army experimental establishment, the Mechanical Warfare Experimental Establishment at Farnborough. Lieutenant Colonel M. S. Brander and Major E. B. Rowcroft, who later became the first Director of the new Corps of Royal Electrical and Mechanical Engineers, both held the appointment of Inspector of Tanks; Major W. S. Tope, the second Director of REME, served as Deputy Assistant Director of Mechanisation in the MGO6 branch and later became the first instructor in mechanical transport duties at the Royal Military College, Sandhurst. Captains F. E. Carpenter, A. E. Damerell and D. R. Hall held appointments at MWEE.

For the time being the Royal Army Service Corps retained its own Inspection branch and Colonel A. D. Owen DSO remained Chief Inspector of Mechanical Transport until his retirement in 1934. In the following year the CIMT was absorbed into the Chief Inspector of Armaments organisation (CIA).

A natural corollary of this move towards centralisation was the creation of a centralised repair organisation for all the Army's transport. But the time was hardly propitious. The rearmament programme was placing heavy demands on available finance resources and the cost of such a radical reorganisation could not be countenanced. And so the Royal Army Service Corps retained its own workshops and repair organisation until, under war time pressures, a new Corps, the Royal Electrical and Mechanical Engineers, was formed, and eventually became responsible for all workshop repairs.

It was fortunate for the Royal Army Service Corps that it was allowed to retain the heavy repair shop, the vehicle reserve depôt and the mechanical transport stores depôt at Feltham during the period leading up to the war. This enabled it to make its own preparations for the equipment of Royal Army Service Corps units for mobilisation, for the storage of reserve vehicles and components, and for the pre-packaging of spare parts.

Policy on the provision of vehicles was reviewed by a War Office committee in 1934 which recommended the principles on which planning should be based. Since the highly successful 6-wheeled vehicle had been adopted as the standard load carrying vehicle, great advances had been made in the design and efficiency of the 4-wheeled vehicle, which now had a greatly improved cross country performance, particularly when it was fitted with large low pressure tyres. Many of the Army's load carrying tasks could be satisfactorily performed by this more economical type of vehicle.

It was therefore accepted that 6-wheeled vehicles would be provided where their high performance was required, but that most of the load carriers for normal tasks should be 4-wheelers.

To ensure that sufficient 6-wheeled vehicles would be available for mobilisation, it was decided to equip all RASC units with this type in peacetime and to make provision for additional numbers to complete war establishment tables, holding them in store on a care and maintenance basis. This also involved the provision of a quantity of special bodies, some of which were held already fitted to the reserve of vehicles in store, and the balance ready for fitting on mobilisation, on vehicles in use with RASC peacetime units.

This placed a heavy burden on the RASC technical units at Feltham, where there was insufficient storage space. Temporary accommodation was taken into use, until a new Vehicle Reserve Depôt could be constructed at Ashchurch, near Tewkesbury, which only became available after the beginning of the war.

Owing to the great demands which the mechanisation programme placed on industry, production of the 6-wheeled vehicles was restricted to five of the firms which had co-operated with the RASC experimental workshops – Messrs Albion, Austin, Karrier, Morris and Thornycroft.

The emergency in 1935 produced a sudden demand for vehicles to equip the units being despatched to Egypt and Palestine, which could not be met from existing stocks. The staff of QMG3 showed considerable resourcefulness in obtaining, at extremely short notice, from the General Motors Corporation, 300 Bedfords, suitably modified for Army use.

The same firm proved equally co-operative when it was decided, after the Munich crisis in 1938, to stockpile a reserve of 500 load carrying vehicles in the Middle East, and again when there was an insufficiency of machine tools for the equipment of workshop lorries to meet mobilisation requirements.

The Director of Supplies and Transport carried out his responsibility, under the Quartermaster General, for the impressment of vehicles and subsidising commercial vehicles, by setting up an enlarged Inspectorate of Supplementary Transport. Captain (Major) R. J. O. Dowse was appointed Chief Inspector in 1935 and was still in the appointment on the outbreak of war in 1939. By 1937 the inspectorate consisted of 35 Inspectors, including the Chief Inspector, with its headquarters in Regents Park Barracks. A number of the Inspectors were re-employed retired officers. The country was divided into four districts, and detailed records of civilian vehicles suitable for military use were compiled. No vehicle was listed unless the inspector judged that it had at least three years useful life. In the event of mobilisation the owners would be instructed to drive their vehicles to selected vehicle collecting centres. In the event, the scheme proved invaluable in the impressment of the planned 14,000 vehicles required on mobilisation. A scheme for hiring vehicles for use in home Commands in replacement for military vehicles on mobilisation also worked smoothly and efficiently.

All these preparatory measures required a great deal of forethought and initiative, and the fact that Royal Army Service Corps units were fully equipped and ready for active service on mobilisation was largely due to the resourcefulness of the small band of dedicated RASC staff officers who, at the time, filled the appointments in the QMG3 branch of the War Office.

Vehicle trials. ABOVE: A modified Mk 2 model P 1925 WD type 500SV 3.5 hp Triumph 'Trusty'. The conversion was carried out in the workshops of P Company RASC and extensive trials were carried out on the machine by Captain C. H. Kuhne RASC. Further tests were carried out on a second machine but were not pursued after 1928 with the advent of the spring frame and cross-country tyres which made the two-wheeled motorcycle more efficient. The 3 wheeled 'Trusty' can be seen at The Museum of Army Transport, Beverley.

ABOVE: Vehicle trials: an Italian *Pavasi* agricultural 4-wheel-drive tractor being tested for heavy gun haulage at Aldershot in 1921. Unfortunately, during the trials it overturned and crushed the driver. LEFT: A French *Latil* logging tractor being tested by P Company RASC, Aldershot in the early 1920s. RIGHT: A six-wheeler vehicle by Vulcan Motors Ltd fitted with pneumatic tyres, to meet the 1924 War Department subsidy scheme 30cwt lorry specifications.

BOVE: To accompany the 30-cwt lorry, a tractor was eeded with equivalent axle load and performance. By tudying the designs of English and Continental tractors, he four-wheeled drive, pneumatic-tyred Hathi was volved with Thornycroft. CENTRE: The Hathi was uilt to WD specification with a Treasury rating of 54 hp. t was fitted with front wheel brakes to increase the load vhen braking — an innovation. BELOW: An assembly of he Hathi and a steam tractor, Maclaren, with trailer, llustrating the two early 1920s schools of thought as to ross country performance: tracked vehicles vs. four- or six-wheeled drive.

MP·8613

22277

OPPOSITE ABOVE: A Renault Dragon carrying a gentler load in Aldershot in 1926, than the machine-gun crew for which it was designed. The passengers are members of the First-Aid Nursing Yeomanry. The RASC were responsible for the maintenance of their vehicles. LEFT: A 1926 experiment in combining six-wheel drive with track, by the weight of the driving wheels running over the shoes. Fitting the track only took a few minutes. RIGHT: A Dragon tested at Aldershot in 1926, after a similar vehicle had successfully crossed the Sahara. The idea behind the rear bogie suspension, was for the torque to be taken up by the frame instead of through the springs. The board over the rear wheels was for carrying a continuous track.

ABOVE LEFT: The floating axle fitted to this 3-ton 6-wheeled Karrier was evolved to give better cross-country peformance. It was not successful. Some 30 of these vehicles were used in India. RIGHT: A 30-cwt Morris Commercial, part of the Royal Corps of Transport Museum Trust collection displayed at the Museum of Transport, Beverley.

OPPOSITE BELOW: Vehicle transporter Garner being tested in 1926. The vehicle mounting the ramp is a one-ton Morris 6-wheeler.

BELOW LEFT: A 30-cwt Morris CD model introduced in 1936. These vehicles replaced the D type and were used in considerable numbers by both the RASC and the RIASC. RIGHT: A 1935 Matchless motor cycle at the Vehicle Repair Depôt, Feltham. This machine was much used by the RASC.

The Sinews of War.

SUPPLIES XX

The Royal Army Service Corps could look back on the 1914-18 War in the knowledge that the supply organisation had met all its commitments without fail. Many were the tributes paid to the efficiency with which the armies had been fed.

Conditions of static trench warfare on the Western Front made the task of supply a comparatively straightforward one, although in other theatres such as the Middle East and East Africa a great deal of improvisation and ingenuity was called for.

Reserve depôts were located at the base and on the lines of communication and these were regularly supplied from the Supply Reserve Depôt in the United Kingdom. Pack trains transported food and forage to the railheads, which were at the rear of the battle zone. From there the divisional supply columns, mostly mechanised, moved supplies forward to refilling points, where rations were transferred to the horse transport companies of the Divisional Train. Local resources were made use of as far as possible, particularly for forage and fresh vegetables.This was the system accepted as standard practice in the immediate post-war period.

During the War, partly due to the drive and energy of Major General Long, the Expeditionary Force Canteen Service had been successfully developed to meet the variety of needs of fighting troops not provided from official sources. In 1916 the canteen service was consolidated under a central organisation, owned and controlled by the Army itself. Thus the Army Canteen Commmittee came into existence, responsible for all Army canteens, worldwide. When the Royal Navy asked to be associated with the organisation in June 1917, the committee assumed the new title of the Navy and Army Canteen Board.

At the end of the war an inter-departmental committee was set up to advise: 'Whether the present system of Navy and Army Canteen Board should be continued without modification or whether (so far as the Army is concerned) it should be brought under War Office administration'. The committee was unanimous in its recommendation that canteens, or institutes, as it was decided to call them, of the three Services should be administered by a joint organisation with the title Navy, Army and Air Force Institute. And so on 1 January 1921 the Navy and Army Canteen Board became the Navy, Army and Air Force Institute.

It was now decided that considerable savings could be made by using this organisation for peacetime distribution of rations. The soldier's ration was divided into three parts; certain staple items such as bread and meat continued to be issued from Royal Army Service Corps supply depôts, or through contracts arranged by local officers i/c supplies; the main bulk of the ration was drawn from the NAAFI; a third element, the messing cash allowance, gave the messing officer a free hand to buy a few extras to add variety to the diet. The total value of a day's ration was less than a shilling and the ration allowance for those not on the messing strength was no more than 1s 3d.

This made it unnecessary to retain more than a few supply depôts; less than a dozen were left worldwide. In small stations, the Royal Army Service Corps items of the ration were obtained from contractors, who delivered direct to units.

Although there were obvious economic advantages in this system, it had two grave effects on the efficiency of the Corps' organisation. First, virtually no supply units existed in peacetime to provide a nucleus on which to expand in an emergency, or in which practical training could be given to the supply branch of the Corps. Secondly, only limited reserve stocks could be held or turned over in the few remaining depôts, and the quantity of reserves was totally inadequate for any major emergency.

The supply depôt at Aldershot retained just sufficient resources to carry out the initial training of supply personnel enlisting in the Corps. Training in the operation of the supply service was limited to theoretical study and the minimal opportunities provided by large scale exercises.

Although the responsibility of the Royal Army Service Corps for the peacetime ration was thus largely curtailed, the Director of Supplies and Transport continued to be responsible for the standard of Army messing, through his inspector of messing, Captain R. G. Leggatt, a retired Royal Army Medical Corps officer, and the training of cooks at the Army School of Cookery at Aldershot, commanded by an RASC officer. At this time unit cooks were regimental soldiers, and not classified as tradesmen; they had little inducement to improve standards. Promotion was limited to the rank of sergeant cook. They were often selected by their regiment for the short sighted reason that they were not very efficient in other aspects of soldiering.

Improvement in the Army's messing was long overdue and a considerable part of the time of QMG6 staff officers was taken up to this end. Major R. T. Snowden-Smith (Deputy Assistant Director of Supplies) and Major T. W. Richardson, who succeeded him in the appointment, were closely concerned with such improvements. Both had the soldier's welfare at heart. A fourth supper meal was provided in the daily diet. A grant of extra messing allowance was made for small messes and the method of assessing individual ration allowances was put on a fairer basis. Improvements were also made in the training and status of cooks and their pay and conditions of service. Even these comparatively small achievements involved protracted negotiation with the financial authorities during the post war period of retrenchment and the financial crisis of the early 1930s.

Such was the situation when the decision was taken in 1934 that preparations would have to be made for a possible war in Europe. The task facing QMG6 was formidable. There were no cadre units from which the needs of an emergency force could be met. The available trained personnel were limited by the restricted peacetime supply commitments of the Royal Army Service Corps; and the reserves held in the depôts were totally inadequate. Any efforts to make good these shortcomings – all of which would be costly – were subject to the delays inherent in peacetime financial control.

One item which cost nothing was the production of a new up to date field service ration scale. It had remained virtually unchanged since 1918. Improved techniques in food preservation and canning and revised nutritional ideas were incorporated in the new scale, including a complete range of tinned or preserved foods, which were a great improvement on the previous hard rations. The Saar Plebiscite in 1935 provided an excellent opportunity for giving this new ration scale a trial. It met with immediate approval and was accepted as the active service ration scale. It was given further trials in subsequent collective training exercises.

In 1932 a new iron ration was introduced, only the third since the South African war, the second having been introduced early in the 1914-18 war. The container was 4 in x 2½ in x 1 in and weighed 6 oz. It bore a label with the words: 'To be consumed only when no other rations of any kind are procurable'.

The ration consisted of sugar, cocoa powder, pea powder, beef powder, cocoa butter and a dash of oil of lemon. The last item was meant to be thirst-quenching.

The building up of reserves was an altogether different matter. It was necessary to persuade the finance branches and the Treasury that reserves were essential, at a time when foodstuffs were so obviously plentiful and easily available in the civilian market. Authority was reluctantly given for the expenditure of a quarter of a million pounds on a thirty days' reserve for 170,000 men.

The situation was slightly eased by the undertaking of the Food Defence Plans Department of the Board of Trade (later to become the Ministry of Food) to meet Service demands in an emergency immediately, without the delays inherent in normal contract procedure.

This was the beginning of a most fruitful liaison between the Army and the Ministry of Food. At the request of the department, a Services liaison officer was appointed to provide a direct link between the Army (QMG 6) and the future Ministry of Food. The officer selected was Colonel J. H. Morris DSO OBE who had retired from the Royal Army Service Corps in 1938, and he remained in the post throughout the war. This was a suitable choice, as Colonel Morris was an experienced Corps officer, who had served in QMG6 as a Deputy Assistant Director from 1924 to 1927. After the end of the 1939-45 war the appointment of head of the Services Supply Branch of the Ministry of Food was filled by Major General H. M. Whitty and Brigadier J. A. Mullington. This continued to be an invaluable contact between the War Office (ST6) and the Ministry of Food.

With the help of the Defence Plans Department, 'dormant' contracts (which only became effective on the declaration of war) were concluded by the Services with many of the principal food manufacturers.

These initial steps paved the way for the establishment of reserves, and ensured that the necessary quantities would be available for an expeditionary force.

These arrangements did not, however, meet the problems likely to arise on mobilisation. The partial mobilisation of the Air Defence of Great Britain in September 1938 helped underline the problem. A strong recommendation was made for a number of supply depôts to be established before the outbreak of war, and agreement was eventually obtained for forty to be set up, where reserves could be held. At first these reserves were restricted to the Royal Army Service Corps items of the ration (meat, biscuits, tea, sugar and salt). Arrangements were made with the Food Defence Plans Department for the delivery of seven days' fresh meat, flour and sugar to each of these forty depôts as soon as they were required.

The Supply Reserve Depôt at Deptford had successfully handled the provision and despatch of supplies during the 1914-18 war. But the depôt was in a most vulnerable location and presented an easy target for an air attack. Although the organisation was fully prepared for expansion in the event of an emergency, space was extremely limited and quite inadequate to allow of sufficient dispersion to safeguard stocks against the fire risk.

A new concrete jetty was built in 1934.

Agreement was obtained in 1936 for a second supply reserve depôt to be established. A site was chosen in south Wales, outside the effective range of enemy aircraft, close to the docks at Barry, where there were excellent rail, road and sea communications and adequate space for dispersion. The completion of this million pound project in the short time available owed much to the untiring efforts of Colonel R. T. Snowden-Smith, who had returned to QMG6 as Assistant Director in 1934. Construction was sufficiently far advanced by the time of the Munich crisis for some of the reserve stocks at Deptford to be hastily moved to Barry.

Another piece of organisation for which Colonel Snowden-Smith could claim responsibility was the creation of the Chief Provision Officer (Food Supplies) – a Colonel, with a Captain to assist him – with the responsibility for controlling the stocks in supply reserve depôts. His duties included the administration of contracts and all the arrangements for overseas shipments. Hitherto these activities had been shared between the War Office (QMG6) and the Supply Reserve Depôt, Deptford.

QMG6's continuing concern over standards of messing resulted in the appointment of a War Office committee in 1936, to enquire into the whole subject of the provision of cooks, their training, pay and conditions of service, and other aspects of catering. The Committee recommended immediate improvements, including special advice on catering in each command, modernisation of cooking equipment and arrangements for the service of meals, more comprehensive training and the enlargement of the School of Cookery, and many other improvements in the status of regimental cooks. Although the formation of a special Corps of cooks was mooted, the committee did not recommend such a development under the prevailing climate.

Many of the recommendations were adopted. In 1938 Sir Isidore Salmon, chairman of J. Lyons & Co, was invited to become Honorary Catering Adviser to the War Office and his appointment had an immediate effect. Specialist catering advisers were brought in from the civilian catering industry. The first of these was Colonel Byford (later Brigadier R. A. A. Byford CBE MVO), who was appointed in 1938 to the Quartermaster General's branch as civilian Chief Inspector of Army Catering. When the Army Catering Corps was formed in 1941, he became its first senior officer as Controller, an appointment which he held until his retirement in 1958.

Steps were taken to modernise cookery equipment throughout the Army, and a new School of Cookery was designed on modern lines; building began in the summer of 1939. A new atmosphere and interest was created which paved the way for the creation of the new Catering Corps.

The provision of sufficient officers and men trained in the handling of supplies remained a problem right up to the outbreak of war, and in many of the hastily formed supply units, officers and non-commissioned officers had to learn as they went along – the inevitable effect of the drastic reduction in the military supply service.

Thanks to the dedication and foresight of the small band of staff officers in QMG6 no disaster occurred on the supply front. The head of QMG6 from 1926 to 1929 was Colonel I. R. S. Shinkwin CMG DSO, who had done three tours of duty in the Supply Reserve Depôt, Deptford, and was an expert supply officer. Later Colonel Snowden-Smith, who had previously been Deputy Assistant Director in the branch, held the appointment from 1934 to 1937, a critical period, when most of the important and far reaching decisions were taken. He had Major T. W. Richardson (later Inspector RASC) as his deputy and Lieutenant (later Colonel) F. W. Price MBE to help him. Major Richardson was an early exponent of the need for an operational ration in a tinned and palatable form, which could be easily handled. He could claim to have initiated the studies from which the 'Compo' ration eventually emerged, a revolutionary concept of supply in the field.

On his appointment as Inspector at the end of 1937, Brigadier Snowden-Smith was succeeded by Colonel H. M. Gale (later Lieutenant General Sir Humfrey M. Gale KBE CB CVO MC), whose team included Major H. M. Whitty (later Inspector), Captain W. H. D. Ritchie (later Director of Supplies and Transport) and the attached officers, Captains D. H. Bond and A. M. Mackintosh, who were concerned with the supply of petrol. These were the officers responsible for the final preparations for mobilisation.

OPPOSITE LEFT: Major General R. T. Snowden-Smith CB CBE MIMechE, as a staff officer in QM6 War Office in the 1930s, was closely involved in improving soldiers' rations. He later became Director of Supplies & Transport, 1940-1943. RIGHT: Classification of Butchers and of Bakers RASC, in 1936. BELOW: Supply Reserve Depôt at Deptford in 1936: preserved meat loaded on the internal railway in the Olympia Shed.

Classification of Butchers.

No.	Technical Description.	Master.	Remarks.
1	Sticking and legging sheep	Thorough knowledge	All privates will be classified in accordance with the standard trade tests published in the Regulations governing the issue of tradesmen's rates of pay, 1923.
2	Skinning and taking bullocks' heads	,,	
3	Skinning briskets	,,	
4	Killing and dressing cattle and sheep	,,	
5	Legging and siding bullocks	,,	
6	Taking bones	,,	
7	Chopping down	,,	
8	Cutting up and preparing meat for issue	,,	
9	Judging live cattle and sheep ...	,,	
10	Judging dead meat	,,	
11	Knowledge of butchery accounts ...	Good	
12	Knowledge of conditions of contract	,,	
13	Supervising and instructing men ...	,,	
14	Erecting and working field abattoirs	,,	

Classification of Bakers.

1	Dough-making	Thorough knowledge	All privates will be classified in accordance with the Standard trade tests published in the Regulations governing the issue of tradesmen's rates of pay, 1923.
2	Moulding	,,	
3	Setting sponges	,,	
4	Scalding dough	,,	
5	Setting	,,	
6	Judging temperature of ovens ...	,,	
7	Baking	,,	
8	Brewing	,,	
9	Stirring ferments	,,	
10	Judging flour, &c.	Good	
11	Rough accounts of receipts and issues, and of bread produced.	To be kept clearly.	
	Field ovens—		
1	Pitching and striking ...	Capable of instructing squads in.	
2	Working		

Supply Reserve Depôt at Deptford in 1936: the oil storage shed.

Petroleum XXI

The provision and delivery of petrol, oil and lubricants during the 1914-18 War presented no particular problems. The quantities involved, compared with those of forage and other supplies, were inconsiderable and even the requirements of aviation fuel for the Royal Flying Corps, for which the Army Service Corps was responsible, called for no special arrangements.

After the end of the war a few mechanical transport enthusiasts foresaw that future large scale mechanisation would create problems and a number of articles appeared in the Royal Army Service Corps *Quarterly*, suggesting that a system of bulk supply of petrol would surely become necessary, and outlining the requirements.

During the first ten years of peace little effort was made to study the problems of petrol supply in the field. Collective training was not resumed until 1924, when the amount required was still minimal. Large scale manoeuvres seldom lasted more than four days. Most units had no difficulty in carrying sufficient fuel in their petrol tanks and unit reserves to make replenishment unnecessary before the end of the exercise. Even as late as 1937, when manoeuvres were designed to last ten days, in order to give a realistic trial of the problems of maintenance of an Infantry division, little attention was paid to the supply of petrol, the main emphasis being placed on a study of the operation of the newly introduced double echelon supply column and the war system for the provision of ammunition and ordnance stores in the field.

It was fully appreciated that the requirement for petrol replenishment was essentially different from that of rations and forage, in that demands were unpredictable and spasmodic, compared with the regular requirement of food and fodder for men and animals. In an emergency, men and horses could continue to operate for limited periods without rations or forage, or on reduced scales. No such flexibility applied to the supply of petrol. Without fuel, mechanical transport became immobile. Petrol supply was more akin to the replenishment of ammunition. Nevertheless, little was done to separate these commodities or to develop different systems for supplies and for petrol. From the War Office down the same staffs and the same types of unit were responsible for dealing with the provision and the distribution of these two different commodities. It required practical experience or realistic field exercises to bring about a realisation that petrol replenishment needed a quite separate organisation from that of supplies.

Eventually, when the Divisional Royal Army Service Corps was reorganised in 1936 and the double echelon supply column was introduced, the divisional baggage company became the divisional petrol company, retaining the commitment of carrying blankets and anti gas clothing, and petrol ceased to be a commodity carried by the supply column.

The provision of petroleum products was of vital importance to all three Services. To begin with, the demands of the Royal Navy and Royal Air Force exceeded those of the Army. It was primarily a national problem, which was handled by a sub-committee of the Committee of

Imperial Defence, the Oil Board, whose responsibility was to ensure that national defence requirements would be met in the event of war. This included the establishment of strategic reserves and plans for meeting the requirements of the three Services. As in other spheres there was a need to reconcile the need for peace time commercial economy with the very different requirements of the Services in time of war. For example, the latter demanded a wide degree of dispersion in the siting of strategic reserves, while commercial practice tended to concentrate stocks in large bulk installations. The availability of material for the Services was inevitably linked to the existing and future trends of the supply and distribution of petroleum products in commercial life. It was, therefore, of the greatest importance that there should be close contact between the Services and the civil oil industry. This was greatly assisted by the appointment in 1935 of Mr William Fraser (later Sir William Fraser CBE) of the Anglo-Iranian Oil Company as honorary adviser to the War Office.

An important decision had to be made on how the distribution of petrol within a theatre of operations should be organised. A War Office committee had been set up to study this particular problem, but it had not come to any firm conclusion.

In 1928 the first serious attempt was made to study the whole problem of petrol supply in the field. A special exercise was designed by the Royal Army Service Corps Training College, to examine all aspects of this increasingly important subject.

At about the same time the Inspector of RASC Services, Major General W. K. Tarver CB CMG, gave a lecture to the officers of Aldershot Command. This dealt with the main sources of supply, the requirements of base petroleum installations in a theatre of operations, distribution from base to railhead, and alternative methods of distribution forward of railhead.

These studies examined the relative merits of bulk supply and distribution in containers. Bulk supply would require a great quantity of expensive equipment and special vehicles. Although the trend of commercial practice was to go over to bulk supply, and the use of cans was rapidly decreasing, the existing petrol tankers in commercial use were unlikely to be suitable for field operations. In the forward area it was considered that bulk supply would be less flexible and more vulnerable than supply in small containers.

General Tarver concluded that in the initial stages, at any rate, distribution would have to be in tins. Peacetime bulk installations would not be suitable for wartime requirements. They were extremely vulnerable and were constructed with an eye to peacetime economy. Plans for locating strategic reserves in the United Kingdom would need to be based on entirely different principles – dispersion, and protection against all forms of attack. Another problem connected with these reserves was that of turnover. The Army's peacetime consumption of petrol was nothing like sufficient to turn over the quantities which it would be required to hold. It would be necessary to have arrangements with the civil industry to deal with this problem.

By 1937 agreed policy envisaged that the main reserves for the Army's requirements would be held in the United Kingdom and that the detailed distribution in an overseas theatre of war would be in the four-gallon container – the 'flimsy' non-returnable can – crated in pairs in wooden boxes. At the same time, wherever possible, reserve holdings and distribution in bulk would be resorted to. This policy involved the construction of plants for the manufacture of containers and the establishment of special reserves for the Services in bulk and in tins, both at home and overseas. In Egypt a reserve of a million gallons in tins was built up in the depôt at Abbassia, on the outskirts of Cairo. Here again, the limited storage life of petroleum products presented a turnover problem.

The shortcomings of the wooden 8-gallon pack, when carried over long distances over rough terrain, with the great wastage incurred, were well known from exercises carried out in the Western Desert in Egypt. As a result, recommendations were submitted for a more robust returnable container. But no suitable replacement was produced until large quantities of captured German Jerricans and their copied allied model were brought into circulation.

This decision, to rely initially on container distribution instead of developing a bulk petroleum system, has been criticised. But, taking into account the need to devote available resources to other high priority rearmament demands, it might well have been wrong to divert valuable resources to an expensive development programme, which might not have had a critical influence on operations. The shipment of petrol in containers to any destination presented no problems, and the ease with which petrol in tins could be dispersed would greatly reduce the vulnerability of stocks. Records show that there was no petrol supply failure, which had any appreciable effect on operations in the early years of the war. By the time when large Allied forces were deployed overseas an elaborate and efficient system of bulk petroleum supply had been developed, which not only simplified rapid replenishment, but also released large quantities of transport for other urgent commitments.

In the four years between 1935 and 1939, great strides were made in preparations for a war which appeared increasingly likely. Bulk reserves for the Services were located in areas considered reasonably safe from attack, both at home and in the Middle East, near Suez. Manufacturing plants for 4-gallon tins were set up at Llandarcy and Avonmouth for the Army, and at Ardrossan and Stanlow for the Royal Air Force. All these arrangements, which were put in hand in close co-operation with the civil oil industry, in particular Shell and the Anglo-Iranian Oil Companies, involved a comparatively small outlay, compared with the cost of equipment required to set up a system of bulk supply.

The staff of QMG6 was woefully shorthanded to deal with this work in addition to its other supply commitments. There were only three officers on the establishment, the senior a colonel. Authority was obtained to attach an additional captain to the staff (later Brigadier D. H. Bond CBE MVO) to be responsible for petrol, with an officer from the Reserve to help him, Captain (later Colonel) A. M. Mackintosh MBE MC. These two officers were chiefly responsible for the preparations made. It is remarkable how much was achieved by such a small staff. Several months later a separate branch – ST2 – was established to deal exclusively with petroleum.

A 4-ton 12-cwt AEC petrol tanker, registered in 1924, and fitted with solid tyres.

Corps Week 1933, an annual reunion of serving and retired members of the Corps, with sporting and other activities, including a tug of war.

Regimental Occasions XXII

The United Royal Army Service Corps Dinner Club was founded as a result of a meeting at the War Office after the armistice by Major General Sir Evan E. Carter KCMG CB MVO. His object was to provide a means of preserving and maintaining the spirit of comradeship which had grown up during the war between Regular, Territorial and temporary officers of the Royal Army Service Corps. The original idea was to form a residential club, with its own premises in London, open to all officers who had served in the Corps.

After a committee had considered various options, it was decided in December 1921 to reconstruct the Club as a dinner club, and the first dinner was held at the Trocadero on Saturday 14 January 1922. Colonel Sir Arthur Holbrook KBE VD DL, the senior temporary officer in the Corps during the war period, was elected President of the Club and, later, in November 1933, in recognition of the great support which he had given to the Club, Colonel E. C. L. Fitzwilliams CMG DL was made Vice-President, an office which he held until his death three years later.

Field Marshal HRH The Duke of Connaught agreed to honour the club by becoming its Patron.

Colonel Sir Arthur Holbrook, who for many years, until May 1936, was Honorary Colonel of the 44th (Home Counties) Divisional RASC (TA), presided at all the club functions until 1936. He died in 1946 aged 96.

For eight years, until 1929, the dinner was held at the Trocadero. It was then decided to hold an annual ball in addition to the dinner. The first was held in February 1930 at the Park Lane Hotel, and in the following year the venue was changed to the Mayfair Hotel and the dinner was held at the Park Lane Hotel, during the week of the Motor Show at Olympia at the end of October.

Owing to the recession and the need for economy the ball was not held after 1931. Three years later it was decided, as an experiment, to move the date of the dinner to the Thursday of Derby week and it was held on 7 June 1934 at the Criterion Restaurant. Out of some 1,000 members there were between 40 and 50 present. In 1936 the dinner was moved back to its former date at the end of October and to a new venue at the Constitutional Club in Northumberland Avenue. This was the last dinner held by the Club. By now eligibility for the Royal Army Service Corps Club had been extended to all officers of the Reserve Forces, and after eighteen years the number of temporary officers of the 1914-18 War who attended was showing a rapid decline. There appeared to be no need for two separate officers' dinner clubs in addition to the annual reunion of the Regimental Association.

The Royal Army Service Corps Club was founded in 1910, largely through the efforts of Major R. Ford (Major General Sir Reginald Ford KCMG CB DSO) who, from the beginning, was the honorary secretary and treasurer, in order to co-ordinate the various recreational activities of Army Service Corps officers and to give financial support to the administration of representative Corps teams. The main function of the Club was an annual 'At-Home' followed by a dinner in London, subsidised from the Club's funds. An annual regimental dinner, an older institution, had been held before the foundation of the Club. From 1910 its funds merged with those of the ASC Club. Up to the outbreak of war in 1914 the cost of a ticket to a member was 5s.

The first general meeting of the Club was held at the RUSI on 28 October 1910. At the first committee meeting Sir Edward Ward was voted into the chair and announced that HRH The Duke of Connaught had consented to become Patron. Sub-committees were appointed for the annual dinner, cricket, the band and the point to point. The first dinner was held at the Hotel Cecil, with Major General H. N. Bunbury presiding.

At the annual general meeting in 1912, Major General Sir H. N. Bunbury KCB, who had been knighted in the Coronation Honours List, was elected President, and four vice Presidents were elected: Major General C. E. Heath CVO CB, Major General C. A. Hadfield, Major General F. T. Clayton CB, and Colonel L. A. Hope CB (as a representative of the retired officers). The dinner in 1912 was held on 12 June at the Savoy Hotel. Major General C. A. Hadfield presided and there were 138 members present out of a membership of 570. The last dinner before the war in 1914 was attended by 129 members.

After the armistice the club was resuscitated and the first postwar dinner was held on Friday 31 October 1919 at the Hyde Park Hotel, when the attendance was 138, including eight major generals. There were only seven subalterns present. The price of a ticket was raised to 10s.

Sir Herbert Bunbury was elected President once more and the four vice presidents were Lieutenant General Clayton and Major Generals Heath, Landon and Koe. Colonel Sir Edward Ward was chairman and the honorary secretary and treasurer was Colonel W. M. Parker. HRH The Duke of Connaught, Colonel of the Royal Army Service Corps, as Patron, showed great interest in its affairs and, when his other engagements permitted, attended and presided at the dinner. He attended the dinner in 1921 and again in 1925, when the Corps Band played for the first time since the war, and in 1927 and 1929.

Sir H. N. Bunbury died early in 1922. No successor as President was appointed for several years. When the Duke of Connaught was not present, Sir Edward Ward, the chairman, presided every year until his death in October 1928. He was a dedicated supporter of the Club and never missed a dinner or meeting.

When Colonel Parker left for India in the summer of 1922, Brigadier General G. S. Dodgson CB CMG succeeded him for a short time as honorary secretary and treasurer. He in turn was followed by Colonel H. G. Burrard CB CSI DSO, who remained in office until after the 1939-45 War.

New Corps Club colours were adopted in November 1921. The tie was dark blue with red stripes.

For ten years, from 1921 to 1931, the dinner was held annually at the Hyde Park Hotel in the middle of June.

Except in the year following the failure of Sir C. R. McGrigor's bank at the end of 1922, which involved a number of RASC accounts, including those of the Club, the price of a dinner ticket was maintained at 10s 6d right up to the outbreak of the 1939-45 War. Members certainly received good value for their money as the menu for 1919 shows.

Toasts were drunk to the King, the Queen, the Prince of Wales, Queen Alexandra and all the Royal Family and to HRH The Duke of Connaught, Colonel of the Royal Army Service Corps. The cost to the Club of this dinner with wines, including champagne, came to less than £3 per head.

Menu Fri 31 Oct 1919

Royal Natives Caviar

Tortue en Tasse

Timbale de Homard Thermidor

Julienne de Sole Murat

Mignonette de Volaille HPH

Pointes d'Asperges au beurre

Selle d'Agneau Samaritaine

Petits Pois Bonne Femme

Pommes Mignon

Sorbet

Faisan en Cocotte Forestière

Salade Japonnaise

Soufflé Hélène

Friandises

Champignons sur Toast

Café

Between the wars the number of members attending the dinner varied between 116 and 146, until 1935. Then there was an annual increase to 185, 195 and the record number of 214 at the Golden Jubilee Dinner in 1938.

Owing to the recession and the drive for economy it was decided not to hold the 'At Home' or the dinner in 1932. For the same reasons the Corps point to point was also cancelled in that year. When the dinner was resumed in the following year, the 'At Home' was not revived.

After the death of Sir Edward Ward, Lieutenant General Sir Frederick Clayton was elected President and Chairman of the Club. From then onwards it became the rule that the senior Colonel Commandant should preside at the annual dinner when the Duke of Connaught was not present. Major General Sir Evan Carter filled this role until he died in 1933, when Major General P. E. F. Hobbs became senior Colonel Commandant.

The venue of the dinner was changed to the Hotel Metropole in 1933 and in the following year was moved to the Junior United Service Club in Charles II Street, where it remained until the end of the 1930s. Owing to the death of King George V in 1936, the date of the dinner was moved from June to November.

At the start, membership of the Club was confined to Regular officers of the RASC. The first change was made in 1931, when it was decided to admit all Territorial Army officers of the RASC as members and to make provision for the election of honorary members (the first two being Lieutenant Colonel H. Niblett CBE DSO and Lieutenant Colonel J. G. L. Pleydell-Nott DSO, both of whom had played such an important part in the research and development of mechanical transport at Aldershot). The Honorary Colonels of Divisional Trains RASC (TA) were invited to become honorary members. This process was completed when in 1936 all RASC Supplementary Reserve officers were admitted to membership. Eligibility for membership was then laid down as follows:
(i) officers of the rank of colonel and above, late Royal Army Service Corps (Regular and Territorial Army); (ii) all officers on the active and retired list of the Royal Army Service Corps of the Regular Army, Supplementary Reserve and Territorial Army (including Officers Training Corps); (iii) regular officers of the Royal Army Service Corps of Commonwealth and Overseas Dominions, and (iv) officers of the Royal Indian Army Service Corps.

Three new Vice Presidents were elected in December 1932: Major General C. A. Hadfield, Major General Sir Llewellyn W. K. Atcherly KCMG CVO (father of the Schneider Cup pilot) and Major General A. E. W. Harman CB DSO, a cavalry officer who served for 12 years in the Royal Army Service Corps (1896 to 1908), and who was at that time GOC lst Division in Aldershot and a great supporter of the Corps. When he retired in February 1934 he was invited to dine in Buller Mess and was afterwards driven home to Wellesly House in the regimental coach by Lieutenant W. O. Phillips and Cpl Degnan. On retirement he was promoted Lieutenant General Sir A. E. W. Harman KCB DSO.

When General Clayton died in December 1933 he was succeeded as President by General Landon, and the Director of Supplies and Transport was elected Vice-Chairman.

An innovation was made in 1935 by inviting the Quartermaster General, Lieutenant General Sir Reginald May KBE CB CMG DSO to become an honorary member, and in the following year the practice of inviting a number of official guests, which had been started in 1933, was continued. The list in 1935 included the Colonel in Chief, the Quartermaster General, Earl Fitzwilliam, the GSO2 RASC Training Centre, the Director of Ordnance Services and the Director of Mechanisation.

The experience of holding the dinner at the end of November instead of in June proved so successful as far as attendance was concerned that this became a permanent arrangement.

General Landon died at the end of October 1937 and he was succeeded as President by Major General S. S. Long CB, the first officer to hold the appointment of Director of Supplies and Transport at the War Office. General Sir Reginald Ford, senior Colonel Commandant, presided at the 1937 dinner. The record accordance of 195 included 24 subalterns, more junior officers than had previously supported these occasions.

The last dinner held before the Second World War was the Golden Jubilee dinner of 1938 which took place on 9 December – the nearest convenient date to 11 December, the anniversary of the birth of the Army Service Corps. General Sir Evan Gibb, the senior Colonel Commandant, presided, and the guests were: the Rt Hon Leslie Hore-Belisha PC MP, Secretary of State for War and Honorary Colonel of the 43rd (Wessex) Divisional RASC (TA), General Sir Reginald May KCB KBE CMG DSO, Quartermaster General; and the Rt Hon the Earl Fitzwilliam KCVO CBE DSO. HRH The Duke of Connaught, the Colonel in Chief, sent a message of regret and disappointment at being unable to be present, on medical advice. A record number of 215 members attended this memorable occasion.

The post war future development of the Royal Army Service Corps Club was foreshadowed in the renewed suggestion in the *RASC Journal* in December 1937 for the formation of a Royal Army Service Corps Institution, on the lines of those of the Royal Artillery and the Royal Engineers. The idea had first been mooted forty years earlier, when it had the lively support of Captain Astley H. Terry, Lieutenant Colonel E. W. D. Ward, Captain P. O. Sandilands and Captain P. E. F. Hobbs. Now it seemed that the concept would once again be stillborn with the imminent approach of the outbreak of war, but it won the support of a future Director of Supplies and Transport, and the Institution of the Royal Army Service Corps was brought to birth by Major General Sir Cecil-Smith, soon after the end of the war and, in due course, became merged with the Royal Army Service Corps Club.

'Corps Week' This annual gathering of the Royal Army Service Corps at its headquarters at Aldershot was inaugurated in 1929, when it was held from Wednesday 22 May to Sunday 26 May. Such was its success that all agreed that it must continue as a yearly event and it was soon extended to the full week.

Major General Davies was then Director of Supplies and Transport and Brigadier J. C. Browne CMG DSO ADC was Assistant Director of Supplies and Transport of Aldershot Command (his rank of Temporary Colonel Commandant on the Staff had been changed to

Temporary Brigadier a year earlier, when the appointment of Brigadier was revived on 1 June 1928). The Commandant of the Training College was Colonel D. C. Cameron, OBE.

The earliest precursor of Corps Week was the first Corps gymkhana held on 12 June 1893, when guests were received by Colonel and Mrs Bridge. In 1903 a function rather similar to Corps Week on a smaller scale was held at Aldershot, when Sergeant Cowan (Captain H. Cowan MC, who for many years was Secretary of the Regimental Association) came second in the obstacle race. Annual sports were also regularly held at Aldershot.

The pattern of Corps Week adopted in 1929 was closely followed in later years and has remained substantially unchanged. It included a cricket match, a rifle meeting, a swimming competition, an other ranks tennis tournament (in which the names of Marrison, Sadler and Ghey appeared prominently), officers' and sergeants' mess dances, an 'At Home' with sports on the Saturday afternoon and the church parade on Sunday.

The salute at the march past after the service was always taken by the Director of Supplies and Transport. The parade, which was formed up on 'W' Square after saluting the war memorial, was addressed for the first five years by Lieutenant General Sir Frederick T. Clayton, as the senior officer of the Corps. After his death in 1933 the Representative Colonel Commandant took over this role. General Sir Reginald Ford introduced an innovation by addressing the parade in 1937, when he was Representative Colonel Commandant, wearing morning dress.

For the first few years the lesson during the service was read by Brigadier W. N. White, first as Commandant of the Training College and then as Inspector of the Royal Army Service Corps. For many years he was churchwarden of St George's Church. When he retired in 1936 the lesson was read, up to the outbreak of war, by the Reverend W. O. Marks DSO, who was a retired RASC officer.

In 1933 a Corps memorial scheme was approved for the recording of the names of RASC officers who had died, on special marble tablets in the chancel of St George's Church. In June of that year the first 57 tablets in memory of those who had died since the South African War were dedicated. These tablets completed the comprehensive collection of Corps memorials in the 'Corps Cathedral'.

Beating Retreat by the RASC Band was first included in the Saturday afternoon programme in 1931, under the direction of Mr C. B. Hewitt MBE, who succeeded Mr C. J. Crosbie as bandmaster in October 1930.

In 1936 a decision was taken that mess dress would be worn at the officers' ball, which was held at the Aldershot Officers' Club – a decision which was not entirely welcome to some senior officers, who considered that their figures showed to better advantage in ordinary evening dress.

To celebrate the Golden Jubilee of the Corps in 1938, special efforts were made to make Corps Week a memorable occasion. The Representative Colonel Commandant, Major General Sir Evan Gibb, wrote to the Secretary of State for War, the Rt Hon Leslie Hore-Belisha, inviting him to the Golden Jubilee church parade on 29 May and asking him to address the Parade on 'W' Square, and to lunch afterwards in the Officers' Mess.

Major Hore-Belisha served as a temporary officer in the Army Service Corps, being commissioned in September 1914. He landed in France early in November and served with the 5th Divisional Train. Later he was at Third Army Headquarters and on the Supply Purchase Board. He also saw service in Salonika and was mentioned in despatches. As Liberal-National Member of Parliament for Devonport he became Financial Secretary to the Treasury and, on 18 November 1933, he succeeded Major General G. F. Davies CB CMG CBE as Honorary Colonel the 43rd (Wessex) Divisional RASC (TA).

The attendance on Corps Sunday in this Jubilee year surpassed all previous records by a long way. The Chaplain General (Reverend E. H. Thorold CB CBE MA DD KHC) preached from the text 'They remembered that God was their strength', and Reverend W. O. Marks read the lesson. The salute outside the church was taken by Major General M. S. Brander, supported by the Inspector, Major General R. T. Snowden-Smith, and the three Assistant Directors at the War Office, Colonels F. W. Beale, E. H. Fitzherbert and H. M. Gale.

When the parade was formed up on 'W' Square, General Gibb, the Representative Colonel Commandant, read a message from the Colonel in Chief:

> 'I congratulate the Royal Army Service Corps on the celebration of the 50th anniversary of their formation. The Corps have every reason to feel a sense of pride in their fine record of service during peace and war, and I am proud to be its Colonel in Chief.
>
> 'I remember the Corps since its formation and its great-grandfather, the Military Train, was commanded by Colonel Clark-Kennedy, who was a personal friend of mine. I hope the Corps may have a very successful celebration and I only wish I could have taken part in some of the week's events. – *Arthur*, Colonel in Chief.'

General Gibb then announced that the address would be given by the Secretary of State for War. During his speech Mr Hore-Belisha expressed his pleasure that: 'His Majesty The King has approved, on this your golden jubilee, the grant of what is in the nature of a present. The Royal Army Service Corps Band, which has hitherto been maintained on a voluntary basis by subscriptions, will henceforth be maintained and recognised officially'.

His stirring speech ended with the words — 'The next war will be a war of industry, but it will also be a war of supply, in which your Corps will once again be not the least vulnerable part of the Army.'

'The Royal Army Service Corps performs for the Army the most human of all its services, for not only are you responsible for its daily bread, but also for the transport of its sick and wounded. Royal Army Service Corps, may I remind you that if this is the age of mechanisation in the Army it is also the age of humanisation? I trust that you will not only practise this principle in your own Corps, but will make all other corps feel grateful to you. I wish you well.'

The eleventh and last pre-war Corps Week was held from 15 to 21 May 1939. All three Colonels Commandant attended the church parade – Generals Macleod (Representative Colonel Commandant), Gibb and Tarver. Two of the officiating chaplains had served in the RASC – the Rev W. O. Marks, who read the lesson, and Rev M. de B. Scott, who gave the address. The latter had served for twenty five years in the Corps until 1927 and had been appointed Honorary Chaplain to the Regimental Association.

Everywhere there was an expectancy in the air and in his address Major General Macleod referred to the great expansion which was taking place in the Regular and the Territorial Armies and to the re-creation of the 'Militia', which would soon place a new and important training commitment on the Royal Army Service Corps.

The Royal Army Service Corps Regimental Association The forerunner of the Association was the Army Service Corps Old Comrades' Club which had been formed in 1909 and which held its first meeting on 24 November of that year and its first dinner in 1910. The honorary secretary and prime mover was Mr J. Lathey, who had been a SQMS in the Army Service Corps. In 1911 it was redesignated The Army Service Corps Old Comrades' Dinner Club and held its third dinner in April 1912 at the Holborn Restaurant, with Colonel Sir Edward Ward in the chair. RSM F. Maclaren, of cricketing fame, was President and Colonel F. Horniblow (Officer in Charge of Records) the honorary treasurer.

In March 1922 the Club, which had been in abeyance since August 1914, was revived as the Royal Army Service Corps Comrades' Dinner Club and held its first post war dinner on 18 October 1922. A number of local branches were formed, notably at Portsmouth, where the first annual dinner of the local branch was held on 11 November 1924, and at Bordon, where a branch of the RASC Old Comrades Association was flourishing in 1925.

At the end of the Great War there was no Regimental Association and no central organisation to maintain contact with the 350,000 officers and men who had served together in the Corps. In the years following demobilisation, when economic difficulties and hardship faced many ex-servicemen, the need for an organisation which could help men who had served in the Corps, or were about to complete their service, became more and more apparent. And so, after discussions at the War Office, it was decided to form a Regimental Association, which formally came into existence on 1 October 1927, with its headquarters in Kensington Barracks.

The objects of the Association were defined as follows:
(i) to foster *esprit de corps*; (ii) to promote the welfare of the Corps and its members; (iii) to organise social gatherings; (iv) to assist members on leaving the Colours to obtain suitable employment at home and abroad; (v) to relieve by means of grants of money members and ex-members and their wives, families and dependents, who are temporarily in necessitous circumstances.

By agreement, the charitable funds which already existed, the RASC Compassionate Fund, which existed as far back as the South African campaign of 1879-1881, and which had been administered since 1906 from the Record Office at Woolwich by the OIC Records with a committee of management, and the RASC Memorial Fund, which was instituted as a result of a conference of officers held in France at the end of hostilities in 1918, together with the RASC Comrades' Dinner Club, were incorporated into the Association, so that all their activities could be co-ordinated centrally. It was also agreed that the Association should take over the management of the RASC Magazines – the RASC *Quarterly* and the *Journal* of the RASC – but this decision was later reversed and the RASC publications were made the responsibility of a committee under the chairmanship of the Commandant of the RASC Training College.

A Regimental Association central fund was established for the administration of the Association and its branches and for the furtherance of its objects. Later the Benefit Fund was formed with the intention of providing a suitable building for the Association and to provide benefits for the members. Field Marshal HRH The Duke of Connaught and Strathearn consented to become Patron, with Lieutenant General Sir Frederick T. Clayton KCB KCMG as Vice-Patron. Major General Sir Evan E. Carter KCMG CB MVO was elected chairman of the committee of management and of the Association. Captain H. Cowan MC became the Association's secretary, an appointment which he held until 1944. As assistant secretary he had Mr W. J. T. Titmas, who remained with the Association until the end of April 1938. The office of honorary treasurer changed hands several times in the first year; at the end of 1929 Colonel F. Falkner OBE TD agreed to take on the office, which he continued to hold until 1939, when he was compelled to give it up through ill-health. He was succeeded by Lieutenant Colonel T. B. Browne OBE, who remained as honorary treasurer until after the Second World War.

The formation of the Regimental Association immediately aroused great enthusiasm throughout the Corps. By May 1928 there were seventeen branches and within a year twenty-five branches had been formed, and the membership was about six thousand. Companies vied with one another in enlisting the highest percentage of members, 5 Company and 35 Company leading the field with 100 per cent membership within six months of the formation of the Association, closely followed by 12 Company in Shanghai.

The headquarters did not stay long in Kensington Barracks, but moved to new premises in Windsor House, Victoria Street, in the summer of 1928.

At one of the last dinners of the RASC Comrades' Dinner Club, the Chairman, Major General Sir Frederick Lander, in his speech proposing the health of the Club, made the prophetic remark: 'If we go on expanding like this, we shall one day finish up by having our reunion in the Albert Hall.'

The first annual dinner and reunion was held at Harrods in November 1928. Over four hundred members were present and General Carter, as Chairman, presided. After this, for the next five years, General Clayton, the Vice-Patron, presided at the dinner, and this practice was followed by his successor as Vice-Patron, Major General Sir Frederick Landon KCMG CB.

On General Carter's death in 1933, General Tarver was elected Chairman. At the second dinner there were 674 members present, the membership had risen to 7,000 and the number of branches had increased to thirty five.

The reunion continued to be held at Harrods until 1933 when, in response to a general request for a less expensive dinner, the venue was changed to Lyons Corner House restaurant, Coventry Street, where the caterers agreed to a price of 4s 6d a head. Unfortunately, after three years, Lyons Corner House ceased to be available owing to a rebuilding programme, and the ninth annual dinner in 1936 was held at the Abercorn Rooms at Liverpool Street Station. This was not very satisfactory as the banqueting room was too small for the numbers attending and it was necessary to use an overflow room. It was decided to look elsewhere for the following year. From 1937 until the outbreak of the war the dinner was held at Thames House Restaurant.

Traditionally, the annual general meeting had been held just before the dinner. This was not always a convenient arrangement and it was agreed that, as an experiment, the AGM would be held in 1937 at Aldershot on the Saturday morning of Corps Week, the day on which the Regimental Association dance took place. This practice was followed for the next few years.

General Landon, who had planned to be present at the dinner in 1937, but had to put off the engagement owing to a sudden illness, died only a few days later, on 26 October 1937. In his absence General Gibb presided, at Thames House.

General Landon was succeeded as Vice-Patron by Major General S. S. Long CB, who had been the first Director of Supplies and Transport. He presided at the last reunion dinner before the war, the Golden Jubilee dinner, at which the Corps Band played for the first time since being recognised as an established Army band, under its new bandmaster, Mr A. Young ARCM.

After hitting a peak of thirty-seven the number of branches in 1937 stood at thirty-six, where it remained until a new branch was formed in 1938 at Newcastle-on-Tyne. Four more branches, in Ceylon, Jersey, Barry and Palestine, brought the number to forty-one by the middle of 1939.

The last annual general meeting before the war was held at Aldershot on 20 May 1939, when the membership stood at 6,790. Major R. H. Frost MBE, one of the veteran members of the Association who had served in the Corps before 1888, put forward a resolution, which was approved, that branch banners should be adopted for use on parades in the open air. There was not time before war broke out for this proposal to be put into effect.

During the war the titles of the senior RASC Association appointments were changed. At the end of 1944 Major General R. T. Snowden-Smith CB CBE was President and Major General E. H. Fitzherbert CBE DSO MC was Vice President. In the same year the retirement of Captain H. Cowan MC as secretary was announced and it was decided to appoint a 'Manager' (changed a year later to 'Controller'). Colonel H. M. Wright was appointed the first Controller in 1944 and Major A. W. Baxter MBE was appointed assistant secretary.

Corps Publications: *The Royal Army Service Corps Journal* Less than three years after the formation of the Army Service Corps in 1888 there was a widespread feeling throughout the Corps that there should be a regimental magazine. The idea was first discussed at Aldershot early in 1891. The first commanding officer of the Army Service Corps at Aldershot, Lieutenant Colonel E. Grattan, was the moving spirit and, thanks to his energy and enthusiasm, the idea soon became a reality, and the first number was published on 1 April 1891.

Colonel Grattan wrote an introduction to the first number under the title 'In Union is Strength'. To quote his words: 'Do we, as a Corps, sufficiently recognise the truth of this axiom?. We think not and for this reason the idea of a Corps Journal was suggested. Excepting India, wherever a British soldier serves, we serve. We are thus scattered over the face of the earth, in bodies of varying strength . . . It is obvious that we want something by which to keep in touch with each other; and after considerable thought and discussion, it was concluded that a Corps Journal, if properly supported, appeared to be likely to furnish the bond needed'. The subsequent history of the *Journal* has shown that this aim has been admirably fulfilled.

The first issue contained sixty pages, 8 inches by 5 inches, and this size was maintained until July 1905, when the pages were enlarged to 11 inches by 8½ inches, which remained unchanged until a pocket size Journal 8½ inches by 5½ inches was produced in September 1944, as a war time expedient.

The Corps motto, 'Nil sine labore', first appeared on the title page of the *Journal* in August 1892, replacing 'For Queen and Country', which had previously been used with the Army Service Corps badge bearing the VR cypher.

The new ASC badge was adopted as a heading in February 1895, but was replaced by a pictorial heading with the words 'The ASC Journal' for the period from January 1906 to April 1909, when it was decided to revert to the ASC badge with a scroll bearing the words 'The Army Service Corps Journal'.

The design of the badge used was changed on 15 March 1913 with the 'ASC' monogram on a white background, and in the issue of November/December 1918 the new title 'The Journal of the Royal Army Service Corps' without the scroll appeared. The badge was still that of the ASC, which was replaced in May/June 1919 by a new G v R badge.

From March 1922 until April 1915 a pictorial heading was once more adopted without a badge, but with the motto 'Nil sine labore'. In December 1925 there was a modification in the G v R badge. The issue for August 1936 carried for the first time the badge with the new Royal Cypher of King Edward VIII, and in the following year the April number showed King George VI's Royal Cypher.

During the period of the Great War, from September/October 1914 to July/August 1919, the *Journal* was produced bi-monthly, and from the October number of 1939 to 1944 it became a quarterly publication. The pocket size edition was also bi-monthly from September 1944, until in October 1947, a full size monthly *Journal*, 9½ inches by 7 inches, with the pre-war cover, was revived.

Apart from changes in size the cover went through a number of changes. From 1926 it was blue with black lettering. Then, in May 1928, the well known white cover with a frame of the Corps colours surrounding the badge and title was adopted, and lasted until after the Second World War. In September 1948, no doubt with the object of giving the *Journal* a more 'trendy' appeal, a pictorial cover was adopted incorporating a photograph of current Corps activities. The new title, *The Waggoner*, was first used in March 1953, with the Corps badge in colour on a white background.

The first editor was Colonel J. A. Boyd CBE. He was succeeded by Colonel Winter. Thereafter the editorship became the responsibility of the chief instructor of the Training Establishment and, later, of a nominated member of the instructional staff. General responsibility, including financial support, for ASC/RASC publications lay with the Corps Club. A RASC publications committee was appointed, with the Commandant of the Training Establishment as President and nominees of the DST, the Corps Club and the Regimental Association as members.

When the *Journal* came of age in 1912, the editor wrote: 'We venture to think, however, that to those who remember it in 1891, the changes effected in twenty one years must seem little short of marvellous. The difference in the organisation, the gradual elimination of the horse in favour of mechanically propelled vehicles, are sufficiently striking examples. We wonder if there will be any horses left when the *Journal* celebrates its jubilee!'

Since the very first number the *Journal* has been an invaluable record of the Corps' activities in all its spheres, professional, sporting and domestic. Few journals can have maintained such a high level of excellence over such a long period or could have called on such a devoted line of editors who, in truth, have held up a mirror to the nature of the Corps.

The Royal Army Service Corps Quarterly There was such a demand for space in the *Journal* for reports and articles on regimental and domestic affairs that contributions on professional matters tended to be crowded out. There appeared to be a definite need for another publication which would be entirely devoted to professional articles. And so, in October 1905 *The Army Service Corps Quarterly* came into being, with Major E. E. Carter (who later became Director of Supplies and Transport) as the first editor. Experience of the South Africa War had revealed the need for a far more professional outlook in the Army. The *Quarterly* was designed to encourage officers to take a greater interest in their professional careers, and to study and discuss the many administrative problems which the Army Service Corps officer was likely to face in war. One feature of the *Quarterly* was the section devoted to the review of other military publications, which aimed at broadening the outlook of the reader and keeping him in touch with recent trends in military thought. The formation of the Army Service Corps' first mechanical transport units provided a rich subject for speculation and study of the probable effects of this revolution in military transport.

Many of the contributors to the *Quarterly* in its early days rose to high rank in the Army Service Corps during the Great War and continued to support it in the post war years. Major A. Crofton Atkins (another future Director of Supplies and Transport) succeeded Major Carter as editor in 1909. Articles covered a great variety of subjects and provide a good insight into the ideas and opinions which were held before 1914.

The *Quarterly* was published regularly from 1905 until July 1914, when the third number of volume 7 was issued. Publication then ceased until the end of the war. The size of the page was 5 ¼ inches by 8¼ inches until July 1910, when it was increased to 6 inches by 9½ inches. When the *Quarterly* was revived in January 1920, mainly through the influence of Major General Sir Evan Carter, who had succeeded Major General Sir Alban Crofton Atkins as Director of Supplies and Transport, the same format was retained until 1925. The first number of Volume 14, which appeared in 1926 measured 8½ inches by 10¾ inches.

Pre war issues of the *Quarterly* had regularly included illustrations, but in its revived form these were sparingly used. A large proportion of the articles appearing in the 1920s covered the experiences of officers in the various theatres of war. There were also discussions, often heated, on the future of the Corps as a result of the 'mechanicalisation' of the Army. Should it become a purely technical Corps or retain its regimental character and traditions? Current developments in mechanical transport and the problems associated with the distribution of petrol were subjects regularly discussed.

Suddenly, without warning, the *Quarterly* ceased publication after the issue of January 1928, the first number of volume 16. No explanation was given, but it must be assumed that it was for financial reasons, as it was acknowledged that the *Quarterly* served a useful purpose. Nearly two years were to elapse before the *Quarterly* was revived in an entirely new form in October 1929.

The new series differed from the earlier publication in a number of ways. In the preface to the first number of this new series it was made clear that the whole expense of publication was being borne by the Royal Army Service Corps Club, and that it had been stipulated that 'the *Quarterly* should deal with the duties of the Corps in modern warfare'. All contributions would be of an entirely unofficial character. Their chief value would be that they were 'the result of recent practical study in the Corps, often in co-operation with other arms'. Henceforth, all the articles were anonymous. Although the opinions were described as 'unofficial', most of them represented the considered views of the staff of the Training Centre, after careful consideration of staff studies and the conclusions drawn from discussions during the new series of war courses initiated. Most of the articles were directly or indirectly concerned with the training and organisation of the Royal Army Service Corps for war.

Although it retained its title as *The Royal Army Service Corps Quarterly* it appeared half yearly and the size was reduced to 9 ¾ inches by 7 ¼ inches. There were no illustrations or advertisements. It was well produced and printed on good quality paper. It was given a wide distribution, both within the Corps and throughout the rest of the Army. In its new form the *Quarterly* owed a great deal to the inspiration and energy of Colonel D. C. Cameron, who became Commandant of the Royal Army Service Corps Training College in September 1929 and was determined to do what he could to improve the readiness for war of the officers of the Corps.

The *Quarterly* soon established a wide reputation throughout the Army and its articles became a recognised authority on Royal Army Service Corps operational doctrine. As a military publication it attracted favourable comment from *The Times* correspondent.

The new series of the *Quarterly* was published for ten years until the summer of 1939, when publication ceased owing to the outbreak of war. During this time twenty numbers were issued in four volumes, which formed an invaluable addition to the official Royal Army Service Corps training manuals.

Although the *Quarterly* was never revived after the 1939-45 War, its traditions were carried on by its successor, *The Royal Army Service Corps* (later *Royal Corps of Transport*) *Review*, which came into being in 1948.

The Corps Band Although there are records of a band of the Royal Waggon Train, based at Mitcham Barracks, Croydon, which was held in high esteem by local inhabitants, the history of the Corps Band really begins with the formation of the Army Service Corps in 1888. The first bandmaster was Staff Sergeant Parr, who joined the Corps from the 5th Royal Irish Lancers. The band consisted of a marching band and a string band and was known as the Army Service Corps Voluntary Band, Aldershot. It had no official recognition and depended for its expenses on subscriptions from the officers of the Corps and income from civilian engagements. When the Army Service Corps Officers' Club was formed in 1910, the officers' subscriptions were taken over by the Club. The bandsmen were found from clerks and horse transport men posted to the establishment of units stationed at Aldershot. Attempts were made to obtain official recognition in 1914 and again in 1919, but without success and it was not until 1938 that this was finally granted.

Staff Sergeant Parr was succeeded by Staff Sergeant Bryce on 29 October 1894, and he remained in charge for nearly ten years. One of his first engagements was to conduct the band at the first guest night to be held in the new officers' mess at Aldershot on 30 January 1895. The band programme survives with the menu, which listed ten full courses.

Three years later, in October 1898, the *Journal* of the ASC records that the bands of the Army Service Corps and the 2nd Battalion, the Northamptonshire Regiment, greeted the arrival of a detachment of the Army Service Corps from Egypt, and that the playing of the regimental march by the ASC Band was much appreciated.

By 1903 Staff Sergeant Bryce had handed over to Bandmaster H. J. Cook, late the Wiltshire Regiment, who made many improvements and continued for twenty years to direct the fortunes of the band.

Among important engagements were the National Rifle Association's meeting at Bisley in 1905, where the band played at lunch during the visit of HM The King, and was complimented by the Colonel of the Regiment, HRH The Duke of Connaught, and Olympia in 1910, when the band played for three days at the Royal Naval and Military Tournament.

The Corps Club continued to support the band with an annual grant, which in 1910 was £235 but, by 1913, had risen to £300, an increase made necessary by the decision to send the band on visits to Army Service Corps stations outside Aldershot.

After the 1914-1918 War determined efforts were made to place the band on a better footing and to make it worthy of a Corps of the size of the Royal Army Service Corps. General Carter, the Director of Supplies and Transport, secured the support of the military members of the Army Council for official recognition, but failed to get the approval of the financial authorities. There were other difficulties. Bandmaster Cook was due for retirement and went in 1922. The failure of McGrigor's Bank, in which all Corps Club funds were invested, forced unexpected economies on the Club's finances and the support that it could give to the band. And it took time to enlist and to train new bandsmen.

After a short interregnum a new bandmaster was appointed in November 1923 – Mr G. J. Crosbie – who had been bandmaster of the 11th Hussars. Under his energetic direction good progress was made in building up an efficient band. In the following year a new blue uniform was adopted for the first time since the war, when the band played at a fête held at Bagshot, and were complimented on their performance by HRH The Duke of Connaught.

The Duke presided at the Corps Club dinner in the summer of 1925, when the Royal Army Service Corps Band played for the first time at this occasion since before the war.

The president of the band committee at this time was the adjutant of the Training Establishment at Aldershot; he was also the honorary secretary and treasurer. In addition the committee consisted of a representative from each of the six home Commands and Northern Ireland District.

With the great improvement in the quality of the band under Bandmaster Crosbie, more civilian engagements were accepted and it was kept busy playing at social and military functions at Aldershot and other Royal Army Service Corps stations.

An innovation in 1926 was a visit to the British Army of the Rhine from 19 to 31 July, which became an annual fixture. This was much appreciated by units stationed there and the following tribute was published in 'Rhine Notes' in the RASC *Journal*:

'Of course, the arrival of the Corps Band has been the cause of providing us with something to talk about for weeks to come. I am not being prejudicially favourable to the band when I say that they have been the cynosure of all eyes in the garrison during the fortnight they have stayed here. In fact, it has been freely said on all sides that they are one of the finest bands heard out here, and moreover, that there isn't a dance band to touch them.'

During the next few years the band's reputation grew and its performances were greatly appreciated in a number of popular resorts. On 17 May 1928 the band played at the Royal Pavilion, Aldershot when HM The King visited the garrison.

In October 1930, after seven strenuous years, during which he had raised the band to a high standard of efficiency, Bandmaster G. J. Crosbie retired and was succeeded by Mr C. B. Hewitt MBE, from the 2nd Battalion, the Buffs, the senior bandmaster in the Army.

The next highlight was in 1933, when the band was first selected to play at the Aldershot Tattoo. Three years later the band made its first broadcast. On 14 August 1936 the band played for an hour on the Northern Regional programme.

Since the establishment of Corps Week in 1929 as a permanent fixture, this was one of the busiest times for the band, with continuous engagements all through the week, including two or three dances, culminating in the church parade on Corps Sunday.

At last, in 1938, the Golden Jubilee year of the Corps, the ambition of the band was realised. The Secretary of State for War announced that the Band would be officially recognised.

Before the decision could be implemented (in October), the band made a most successful first visit to Northern Ireland at the end of August, during which a broadcast was made from Belfast and a concert was given at Belfast Castle, the residence of the Earl of Shaftesbury.

This was Mr Hewitt's last series of engagements and, on his retirement, the first official bandmaster was appointed – Mr A. Young, ARCM, who had previously been bandmaster of the Duke of Cornwall's Light Infantry. The approved establishment of the band was one Bandmaster, one Band Sergeant, one Band Corporal and 22 Bandsmen.

An increase to this establishment was approved at the outbreak of war. It then consisted of one Bandmaster, one Band Sergeant Major, one Band Corporal (local Sergeant) and 33 Bandsmen.

During the war the band was kept busy, doing its best to keep morale high and covering many miles of travel. Mr Young, who had contributed so much to its efficiency during its formative years as an official band, left in January 1944 to become Director of Music, Royal Engineers, and was succeeded by Bandmaster Pope, who had been bandmaster at the Royal Military Academy, Sandhurst. He found himself immediately plunged into overseas tours of Italy, Sicily and North Africa, which occupied the six months from March to September 1944.

Before the end of the tour Mr Pope was recalled to take up a new appointment. When the band returned home on 14 September 1944, Bandmaster Pope was succeeded by Bandmaster J. F. Dean ARCM psm, who also came from the Duke of Cornwall's Light Infantry.

Two engagements of the band brought feelings of nostalgia to many members of the Corps. In January 1945 the RASC Headquarters' Mess celebrated its 50th anniversary and the band played, at the guest night, selections of the music which had been played at the first guest night held in Buller Officers' Mess fifty years before. In 1947 the band played once more at the Royal Tournament at Olympia, where it had last played in 1910 for three days. This time it played for the whole period of seventeen days.

December 1947 marked a most important step in the history of the band when it became a minor staff band, entitling it to a Director of Music, and Bandmaster J. F. Dean was promoted to the rank of lieutenant (Director of Music). The other change involved an increase in establishment, which now became: one Director of Music, one Band Sergeant, two Band Corporals, two Band Lance Corporals, 30 Bandsmen and 10 Boy Musicians.

All NCOs in the band held local ranks above those authorised by the establishment, to compensate for the comparatively slow rate of promotion.

The administration of the band from now onwards was in the hands of a strong committee consisting of the Band President – Commandant, RASC Training Centre, and the Band Committee – Deputy Commandant, RASC Training Centre; OC No 1 Training Battalion RASC Aldershot; Adjutant No 1 Training Battalion RASC Aldershot (ex officio Secretary and Treasurer).

In 1956, the RASC Band received the status of a major staff band with a Director of Music in the rank of captain or above. Major J. F. Dean MBE continued as Director of Music until his death in December 1962.

The Regimental March — *Wait for the Wagon* The following is the published account of the adoption of *Wait for the Wagon* as the Regimental March of the Army Service Corps in *Fifty Years of Army Music* by Lieutentant Colonel J. Mackenzie-Rogan.

'The Sultan of Zanzibar was visiting this country in 1875, and one of the ceremonies arranged for him was a review on a large scale in Long Valley, Aldershot. All available men were ordered to be on parade. It was a bright warm day, and we had the usual clouds of dust in the famous valley. Field Marshal the Duke of Cambridge was present, and after the Royal Salute, he and the Sultan inspected the troops.

'Here I may relate an incident which has, I think, some historical significance for the ASC, now the RASC. It had not been customary, as far as I am aware, for the ASC to march past at reviews, but on this day the Duke was asked by his Chief of Staff if they should do so. He replied that they were to do so, and that the Cavalry bands should play for them as they came by. The Chief of Staff asked what tune should be used for their march past. The Duke replied: "Tell the bands to play *Wait for the Wagon*". And that, to this day, I understand, is the march of the RASC.'

This march continued in use until 1945, although some time after 1900 the march also included some sixteen bars of *Begone, dull care*, now the regimental march of the Royal Corps of Signals.

By 1945 it was generally agreed that it was too short and its continued repetition too monotonous. General Kerr, Director of Supplies and Transport, set up a committee at Aldershot to revise and improve it. The Assistant Commandant of the RASC Training Centre, Colonel C. J. Williams and Mr J. F. Dean, the bandmaster, were the leading members of this committee.

In due course a new march emerged, in which a South African folk tune, the *Trek Song*, was added to the original *Wait for the Wagon*. The new arrangement, after approval by the Royal Military School of Music, was played at Kensington Palace Barracks by the band on VJ-Day, 15 August 1945 and received an enthusiastic reception from those present.

After the Colonel in Chief, HRH The Duke of Gloucester, had given his approval, the new march was notified in Army order number 36 of 1946, the original title *Wait for the Wagon* being retained. The march consists of three parts: the Regimental Call, the *Trek Song* and the original *Wait for the Wagon*.

Neither the composer of *Wait for the Wagon* nor the lyricist is known with certainty. Records at the music room of the British Museum attribute the music to George F. Knauff and suggest a date of composition of about 1858. There is, however, another claimant. A book, *Our family songs and those who made them*, by Henry Holt, was published in New York in 1881, which stated that the music was composed by R. Bishop Buckley, born in England about 1810; he emigrated to the United States, where he organised Buckley's Minstrels in 1843. The balance of evidence favours Knauff.

The origin of the words is even less certain. They appeared in the *Song Book*, published by Macmillan and Co in 1866 as an 'old English folk tune'. The song was also included in a volume of American songs published in Baltimore, USA, in 1840. The words in the *Song Book* are:

'Will you come with me, my Phyllis dear,
 To yon blue mountain free,
Where blossoms smell the sweetest?
 Come rove along with me.

It's every Sunday morning, when I am by your side,
We'll jump into the wagon, and all take a ride.

(Refrain) Wait for the wagon,
Wait for the wagon,
Wait for the wagon,
and we'll all take a ride.

'Where the river runs like silver,
And the birds they sing so sweet,
I have a cabin, Phyllis,
And something good to eat,
Come listen to my story, I will relieve my heart,
So jump into the wagon, and off we will start.
(Refrain)

'Do you believe, my Phyllis dear,
Old Mike with all his wealth,
Can make you half as happy
As I with youth and health?
We'll have a little farm, a horse, a pig, a cow,
And you will mind the dairy, while I will guide the plough.
(Refrain)

'Your lips are red as poppies,
Your hair so silk and neat,
All braided up with dahlias
And hollyhocks so sweet.
It's every Sunday morning, when I am by your side,
We'll jump into the wagon, and all take a ride.
(Refrain)

'Together on life's journey
We'll travel till we stop,
And if we have no trouble,
We'll reach the happy top.
Then come with me, sweet Phyllis, my dear, my lovely bride,
We'll jump into the wagon, and all take a ride.
(Refrain)'

Two other 'modern' versions have been written, one by Rev M. de B. Scott (who became the Honorary Chaplain to the Regimental Association, after serving for 25 years in the Corps), which appeared in the ASC *Journal* of November 1906; and another by Major J. C. Mackie RASC, which was published in the *Journal* of May 1933. All three versions were reprinted in *The Waggoner* of May 1960.

The words of the *Trek Song, (Vat Jon Goed en trek Ferreira)*, translated from the Afrikaans, are:

'Pack your things and trek, Ferreira,
Johnny with a wooden leg.
Pack your things and trek, Ferreira,
Johnny with a wooden leg.

Turn, turn all on the one side,
Turn, turn all on the one side,
Turn, turn all on the one side,
Johnny with a wooden leg.

General Buckle has carried out extensive research in South Africa into the origin of the tune and the words and summarised his findings in *The Waggoner* of July 1974. He was convinced that they were written and composed in the Eastern Cape about 1870. According to his account, Miss Ada van Aardt wrote the words and Mrs Maloney, her music governess, composed the music. It was said to have been sung to discourage a persistent but unwanted lame suitor, Jannie Ferreira.

The song became popular all over South Africa and, during the early days of the South African War, was used as a drinking song by Boer Commandos. General Buckle concluded: 'Whatever its origins and uses through the years we have, as part of our regimental march, what Laurence Green (a well-known Cape author) describes as "the most famous of all the old Cape Town songs". '

The Corps Fanfare The Corps Fanfare, *Fanfare for an RASC Occasion*, was specially written and composed for the 1948 Corps Week by the Director of Music, Lt J. F. Dean. It was first played in St George's Church on Corps Week Sunday on the unveiling of the memorial to officers and men of the Royal Army Service Corps who lost their lives in the Second World War, 1939-45.

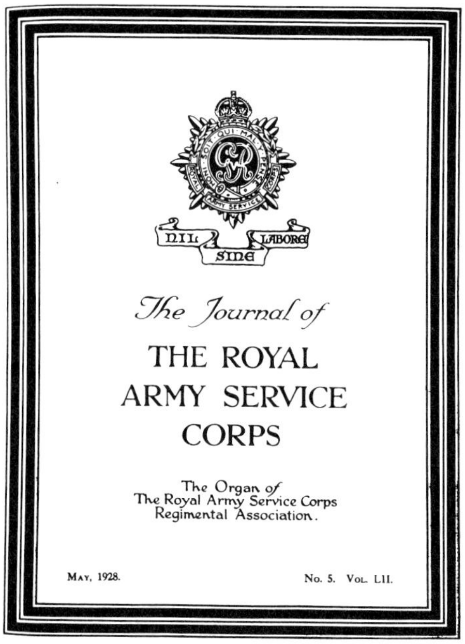

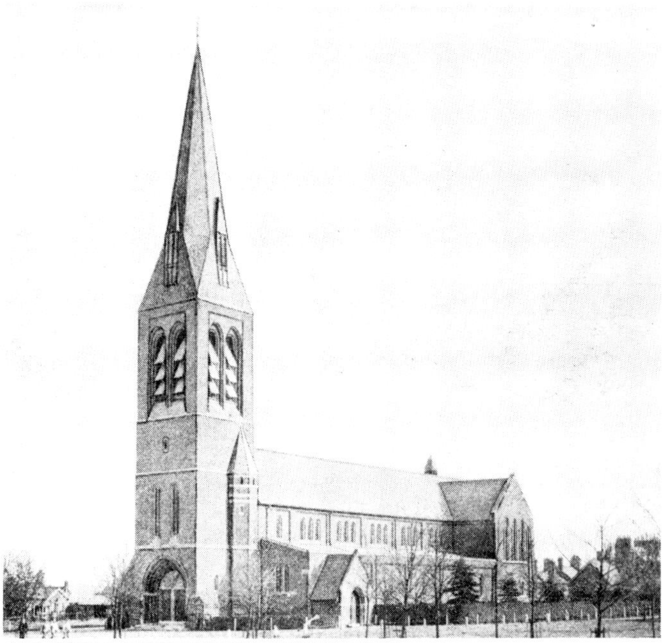

LEFT: The 1928 cover of the Corps *Journal*. RIGHT: The Corps Church, St George, now renamed St Michael & St George. BELOW: The Rt Hon Leslie Hore-Belisha, Secretary of State for War, taking the salute on Corps Sunday, 29 May 1938.

The RASC Band in 1934, Bandmaster Mr C. E. Hewitt MBE.

Pipes and Drums 51st (Highland) Division RASC TA at Nairn in 1937, the first official Pipes and Drums in the RASC. The pipers are wearing the MacDuff tartan.

LEFT: The Corps' War Memorial in Buller Barracks, Aldershot; in both World Wars just under 10,000 officers and soldiers of the Corps were killed in action. RIGHT: Major General S. S. Long CB, President, RASC Club 1937; vice-President RASC Regimental Association, 1938. BELOW: Colonels Commandant, Corps Sunday 1939 — L to R: Major General C. W. Macleod CB CMG DSO; Major General Sir Evan Gibb KBE CB CMG DSO; Major General W. K. Tarver CB CMG.

Sports and Games XXIII

Between the two wars there was hardly a game or sport which was officially recognised by the Army in which Royal Army Service Corps players failed to make their mark. Two years stand out as memorable for the distinctions gained by the Corps – 1923 and 1937 — when the Corps won the Army Football Cup.

1923 was a veritable 'annus mirabilis' for the Royal Army Service Corps in the field of sport. While the Army Football Cup was being won at Aldershot, Woolwich was distinguishing itself in mounted sports, our swimmers were winning laurels in Constantinople and the Hellespont, where Captain M. N. Dalton, with Sergeant W. Allen, Corporal Johnson and four drivers, swam the straits in 80 minutes, emulating the classical feat of Leander, which was repeated by Lord Byron and Lieutenant Ekenhead in 1810; RASC golfers were winning championships in Turkey, Ceylon, India and Egypt. Internationals from the Corps were playing Association and Rugby football and Lieutenant Hegan was captain of the Army football team. On the ranges at Aldershot the Corps had one of its most successful seasons. Cricket was flourishing at home and abroad and at Peshawar most of the outdoor sports were administered by RASC officers. At one time it was fashionable to criticise the larger corps in the Army for 'packing' their teams by means of selective postings to particular stations. But so scattered was the sporting talent of the Corps at this time, there was no justification for such a charge against the Royal Army Service Corps. The standard of cricket was maintained at Aldershot at the same time as names like Fitzherbert, Tudor and Marrison were prominent on cricket fields in Peshawar and on the North West Frontier.

In 1937, coronation year, seven of the team which won the Army Football Cup played for the Army, and Corporal S. Waite was selected to play for Wales. The Feltham tug-of-war teams were at the height of their successes, 130-stone Army Champions for the fifth consecutive year, 110-stone Army and inter-service champions, and 100-stone AAA champions for the fourth successive year. In boxing, Feltham won the Noble Trophy for the second time and came close to winning the King's Trophy; H. C. D. Huxham created a record by winning his ninth consecutive Army championship. Lieutenant J. March won the Army light-weight title, and Corporal Fisher and Driver Allday boxed for the Army. At golf the Royal Army Service Corps Training Centre team won the Army inter-unit championship for the first time by a record margin. Under the captaincy of Major A. C. Booth MC, the Training Centre team won the Duke of Connaught Cup for the champion revolver team, and Trumpet Major Townsend won the Army Rifle Association Revolver Challenge Cup. At swimming, Service Companies, Aldershot, won the relay race in the Army championships and Private Ford won the fancy diving title. Captain J. R. Allen came second on Hilton in the King George V Challenge Trophy at the International Horse Show. For the first time the Corps entered a team for the modern pentathlon, which had been started in 1929. The team was placed fourth, with Lieutenant J. March coming third in the individual placings. At the end of the year the Royal Army Service Corps received the personal congratulations of the Adjutant General (General Sir Harry Knox) on this remarkable run of sporting successes.

Association Football When the war at last came to an end in 1918 the Royal Army Service Corps was still holder of the Army Football Cup, won by Woolwich in 1914, the first time the Corps had won the cup, although it had reached the final in 1899.

There had been several prominent Corps names in Army football before the First World War. Ford (Major General Sir Reginald Ford KCMG CB DSO), White (Brigadier W. N. White CB DSO), Hushar (Lieutenant Colonel W. S. Hushar), Gibb (Major W. Gibb MBE) and Cowan (Captain H. Cowan MC), who for many years was secretary of the Regimental Association, were all distinguished players. General Ford and Brigadier White both served for many years on the council and committee of the Army Football Association. Brigadier White captained the Army team in 1902-3. Colonel Hushar played for the Army from 1911 to 1922 and was captain on a number of occasions. He was also a regular member of the Hampshire team. Both Hushar and Gibb played in the winning side in 1914.

The competition was re-started in the 1919-20 season, when thirteen RASC teams entered for the Army Cup. Two years later, when the Corps was reduced to its peacetime strength, the RASC was split into ten divisions for entry to the competition.

In the fourth season after the war, 1922-23, No 1 Corps Depôt RASC, Aldershot, was successful in winning the Army Cup for the second time. Lieutenant Hushar, who captained the team, and CSM Cooper were the only two players from the 1914 winning team. Lieutenant F. W. Clarke scored the winning goal, when the 2nd Battalion The Cameron Highlanders was beaten 1-0 after extra time. The final was played on the Aldershot Command Central ground in the presence of the King and Queen, accompanied by the Prince of Wales and Prince Henry. In addition to Husher and Clarke there were two other officers in the team: Lieutenant K. E. Hegan and Lieutenant G. B. McDonald, and three other Army players, the goalkeeper, Cpl Wood (SQMS W. E. Wood), CSM Cooper and Sergeant McAvoy (CQMS R. D. McAvoy).

Lieutenant Hegan had transferred to the Corps from the Dublin Fusiliers in 1922 and was already an amateur international and Army player. It is generally accepted that he was by far the greatest football player in the RASC. He won three full international caps for England and played in more than twice as many amateur international matches than any other Army player. He represented the Army on innumerable occasions and was captain of the Army team from 1923 to 1929. Later he served on the committee of the Army Football Association and was a member of the Army team selection committee. General Fitzherbert, who was well qualifed to judge, wrote that he had 'every right to be hailed as the greatest player the Army has ever produced, and probably the finest amateur outside left that has ever lived.' When he finally retired from football he did not lose the accuracy of his kicking, as the following incident shows. In September 1938, while the Munich crisis was at its height, the author and a few others were discussing football with Major Hegan in a desert camp not far from Mersa Matruh. Major Hegan wished to give a practical demonstration of the point he was emphasising, but there was no football available. The author produced his tooth glass, which was made of 'unbreakable' glass. Hegan, with a carefully aimed kick, sent it flying and it exploded into a thousand pieces.

Between the two wars the Royal Army Service Corps had the unique distinction of providing an international football player for each of the four home countries. Captain K. E. Hegan played for England on more than twenty occasions, Captain T. M. R. Briggs for Scotland against England in 1928, Cpl Vance for Ireland against England in 1925, and Cpl S. Waite for Wales against England in 1937.

In addition, the RASC provided many other Army footballers, including Sgt F. Twine, Cpl Spittle (CSM B. J. Spittle), Sgt Mallon, who captained the Army team in 1929, Sgt A. Bolton, Cpl H. G. Bowtle, LCpl J. Dallas, Pte L. A. Sadler and Pte Wallace. Apart from the Army Cup, the RASC Training Centre entered for a number of other league and cup competitions open to Aldershot Command and Hampshire teams. Among other successes they won the Aldershot Command senior league in 1931, 1934 and 1935.

Overseas the Royal Army Service Corps Cairo XI won the Egypt Command Cup in 1932, defeating the Royal Engineers in the final, and they won the cup again in 1939, becoming joint holders of the Command Inter-Service Shield, after a drawn game with the Royal Air Force.

On Easter Monday, 29 March 1937 the RASC Training Centre won the Army Cup for the third time, beating the Training Battalion Royal Engineers 2 goals to 1. It is doubtful whether any winner of the Army Cup has had such an impressive goal average on the way to the final. In seven rounds, 34 goals were scored by the winners with only 3 goals scored against them. In this coronation year cup final, no less than seven members of the Corps team were Army players – Cpl S. G. Waite, Cpl L. A. J. Sadler, LCpl B. Watson and Pte A. G. Davies of the forwards, LCpl J. D. Gray and Cpl A. G. Stewart at half-back and Cpl J. Dallas at back.

Although the Training Centre team failed to retain the Army Cup in 1938, it had the consolation of winning the Hampshire Senior Cup and the Aldershot Senior Cup, and it had four representatives in the Army team, LSgts Stewart and Owen, Cpl Waite and LCpl Gray. In 1939 Cpl Gray was the recognised captain of the Army football team.

The Army Football Association Challenge Cup Trophy is a silver replica of the famous Warwick Vase, made in 1838 by Paul Storr, a well-known silversmith, and bought by the Army Football Association in 1888, the year in which the Army Service Corps came into existence.

The tradition of football in the Royal Army Service Corps survived the Second World War and further successes brought the Army Cup back again to Buller Barracks.

Cricket Colonel E. Grattan, who commanded the Army Service Corps at Aldershot in 1890, was responsible for creating, with the assistance of Captains F. C. A. Gilpin, F. Horniblow and SSM Rose, the recreation ground in front of the officers' mess.

A representative Corps cricket side first played in 1908, under the captaincy of R. B. Airey (Colonel R. B. Airey CMG DSO), and such well-known names as F. A. McLaren (Captain F. A. McLaren MBE), W. N. White (Brigadier W. N. White CB DSO), E. H. Fitzherbert (Major General E. H. Fitzherbert CBE DSO MC) and W. K. Tarver, (Major General W. K. Tarver CB CMG) were among the players.

Colonel Airey, who found himself serving in Aldershot after the war, was determined to revive Corps cricket as quickly as possible, and a full fixture list was arranged for the 1920 season. A nucleus of first-class players survived the war years and, with a plentiful supply of up and coming young players, Corps cricket soon regained the high reputation of pre-war days.

In a series of articles entitled 'A Cavalcade of Corps Cricket', in the RASC *Journal* during 1936, Colonel Fitzherbert, who succeeded Brigadier White as captain in 1934, had no doubt that Brigadier W. N. White was the best cricketer who had ever played for the Corps. He was one of Hampshire's best batsmen and he played regularly for the Army. He captained the Corps side for more seasons than anyone else from 1921 to 1933. He made more runs than any other player and in 1927 he made 762 runs in Corps matches. He was also a first-class footballer and captained the Army team for two years.

The Royal Army Service Corps was not short of other first-class players. Apart from E. H. Fitzherbert, whose playing career extended from 1906, when he was on his junior officers' course, to the late 1930s, C. L. St J. Tudor (Brigadier C. L. St J. Tudor CBE MC), who was also an Army hockey player, played for Sussex and captained the Army at cricket in 1926 and 1927, E. D. Dynes (Brigadier E. D. Dynes CBE) was captain of the Bedfordshire XI and was a regular Army player. He was one of the Corp's best all-rounders. Major C. W. Edwards (Col C. W. Edwards DSO) and Major A. U. Udal were both first-class county and Army players. P. S. Whitcombe (Major General P. S. Whitcombe CB OBE) was a first class batsman, who played for Essex and the Army, and T. E. Clarke (Colonel T. E. Clarke OBE MC) was an excellent all-rounder.

In General Fitzherbert's opinion, the three greatest bowlers were F. McLaren (Captain F. A. McLaren MBE), a Hampshire county player, who played for the Corps until 1929, H. M. Hinde (Brigadier H. M. Hinde CB CBE) and F. Marrison (Major F. Marrison MBE). Between 1929 and 1935 Marrison took 259 wickets in Corps matches and in 1936, against the Royal Corps of Signals, had the remarkable bowling analysis: 14.4 overs; 13 maidens; 1 run; 7 wickets: 88 balls bowled; average per wicket 0.14.

Among the many footballers who were prominent in the Corps cricket side were: F. W. Clarke (Colonel F. W. Clarke OBE), Cpl L. Sadler (Major L. Sadler), W. Gibb (Major W. Gibb MBE) and Cpl Hamlen (Major F. W. Hamlen), who played full-back for the Army Rugby football team. Both the last two were first-class wicket-keepers. Between 1925 and 1935 there was only one season (1926) in which the Corps lost more matches than it won. Considering the strength of the fixture list during these years this is a good indication of the consistently high level of Corps cricket during the captaincy and presidency of Brigadier W. N. White.

While a high standard was maintained at Aldershot, the game was flourishing in other Corps stations at home and abroad. In Cairo the RASC team won the command cricket competition against strong opposition in 1926 and reached the final again in 1934 and 1939. While the Corps was still serving in India the names of Corps cricketers were prominent. In Palestine, in spite of the troubles and the uncertainties of some of the pitches, RASC cricketers, led by Major P. A. Arden, lived up to the standards of the Buller ground.

1937 was another good year, with six matches won, one lost and one drawn, Captain Dynes heading both the batting and bowling averages.

In 1938, Major W. Gibb MBE and Lieutenant F. Marrison celebrated their 28th year as Corps cricketers. They played their first game for the Corps as lance corporals in 1911. Gibb was selected as wicket keeper and Marrison for his bowling. As tail end batsmen, Gibb scored 43 and Marrison 37 runs to help their side to victory. Throughout this period Gibb was the Corp's most successful wicket keeper.Marrison's record of 2,769 runs scored, 526 wickets taken and 103 catches held is one unlikely to be overtaken.

In the same season, his last before being posted to Singapore, Sgt L. A. J. Sadler scored a record 183 runs against the Free Foresters, the highest score in Corps cricket matches. The previous record had been held by Brigadier W. N. White with 174 against the Royal Tank Corps in 1927.

A number of high-class younger players were coming to the fore as the 1930s drew to a close. The 1938 team still included some of the old guard like Major P. S. Whitcombe, Major F. W. Clarke, Major W. Gibb and Lieutenant F. Marrison, but more than half the team belonged to a younger generation. Lieutenant R. G. J. Kingsmill, Lieutenant A. Macdonald, Second Lieutenant W. M. E. White, Second Lieutenant H. F. J. Wigram and Second Lieutenant J. L. Chambers had already made their mark in Corps cricket by the time that war was declared in 1939.

The reputation that the Buller ground has maintained as one of the best service cricket fields owed a great deal to the timeless efforts of Mr Jackman, who for 37 years from the early 1920s was its devoted groundsman.

Rugby Football and Hockey Royal Army Service Corps rugby and hockey teams have not had successes comparable with those of the Association Football sides. Nevertheless, in both games Corps teams reached the final of the Army competition – the Army Service Corps

Aldershot Hockey team in 1910, and No 1 Corps Depôt, Aldershot rugby team in 1923, when they lost to the Welsh Guards.

Two outstanding Rugby football players were J. A. Middleton and H. M. Hinde. Both were regular Army players from 1921 onwards and Middleton captained the Army team in 1929. In 1922 he was selected to play full back for England against Scotland. Captain Hinde was selected for an English Rugby trial in 1923, and was perhaps unlucky not to get his international cap. For many years these two players were an inspiration to RASC Rugby, wherever they served, at home or abroad. The Welsh international, Lieutenant B. E. W. McCall, served for a short time in the Royal Army Service Corps and helped the RASC Training Centre 7-a-side team to reach the final of the Aldershot Command competition in 1937.

Corporal Hamlen (Major F. W. Hamlen), a first class all round games player, represented the Army on a number of occasions in the mid 1930s. A few years later, in 1937 and 1938, Lieutenant J. D. Hennell played for the Army. In 1939 Lieutenant Chambers represented the Army against the Royal Air Force at scrum half.

Ever since they played against each other in 1924, when they were on their works courses in Glasgow, Lieutenant Smith (Major General Sir Cecil Smith), Albion Motor Car Company Limited, and Lieutenant Goodfellow (Major General H. C. Goodfellow), Halley's Industrial Motors Limited, have both been ardent supporters of Rugby football in the Corps and have contributed greatly to the success of the game.

Among individual hockey players S. Acheson, G. K. Archibald, E. H. Fitzherbert, W. N. White, C. W. Edwards, J. C. L. Godfray, S. S. Saulez and R. T. Snowden-Smith of the older generation, all represented the Army in the early 1920s and earlier. Acheson also played for the Combined Services in 1922. Godfray captained the Army team in 1909. Later J. D. F. Fisher was an English international in 1930 and J. A. Mullington was selected to play for the Anglo Irish hockey team against the Rest of Ireland in 1930.

With high class players like C. L. St J. Tudor and F. S. Clover, the Corps maintained a high standard against first class opposition. With the influx of a number of university players in the mid and late 1930s the prospects in the Army Cup competition looked distinctly promising. Among these young players were Lieutenant J. F. Rush, Lieutenant R. F. Marsh, a Berkshire county player, who played for the Army and the Combined Services, and Second Lieutenant Bullough, who represented Essex. Corporal Rogers, who played in the Corps side in 1937 and 1938, was one of the best backs who played for the Royal Army Service Corps. The outbreak of war in 1939 put a temporary end to further progress in building up a winning team.

Lawn Tennis Colonel A. Berger was the best lawn tennis player who has served in the Royal Army Service Corps. He was Army Rackets Champion in 1909 and, with Major Puckle, he won the Army rackets doubles championship in 1910, and 1911 and the Army lawn tennis doubles championships in 1910, the first year in which the championships were held. In 1923 he reached the fifth round of the singles championship at Wimbledon and in the same year was selected to play for England in the International match against Scotland. He was probably the best Army player who never won the Army singles championship. He very nearly did so in 1923, when he narrowly lost the final in the fifth set. He won the Army open doubles championship again in 1929 with Lieutenant A. T. Grimley, the Dorset Regiment. When he retired in December 1929, he became a lawn tennis professional and had a number of successes in the professional championships of Great Britain.

Among Berger's contemporaries there were three other outstanding lawn tennis players, A. St J. Mahony, J. D. F. Fisher and G. Bright Holmes. Captain Mahony played for Ireland in the Davis Cup in 1927. Between 1923 and 1926 he and Lieutenant Bright Holmes monopolised the Army singles championships in Egypt, Mahony winning in 1923, 1925 and

1926, and Bright Holmes in 1924. Together in 1926 they won the Army doubles title and the open doubles championship of Egypt. Bright Holmes, who was also no mean cricketer, won the singles in the Cairo open lawn tennis tournament in 1925. After they were posted home it was twelve years before another member of the Corps won the Army singles championship in Egypt, when Lieutenant Turpin won it in 1938.

J. D. F. Fisher was one of the few players in the Corps who succeeded in beating Berger, which he did in the final of the RASC championship in 1926. In 1925 he was one of three RASC players who reached the last eight in the Army championships, the other two being Berger and Price. With Captain J. H. Frowen, Royal Artillery, he won the Army doubles championship in 1930. The next year he transferred to the Indian Army Service Corps and was lost to RASC tennis, except when he came home on long leave.

Captain W. C. Price played for the Army in 1927 and won the Army open doubles with Captain A. E. Browne, Indian Army.

In the events restricted to other ranks in the Army championships, Baxter (1st class SSM A. V. Baxter) won the singles title in 1926 and 1927, and with his brother (1st class SSM A. W. Baxter) won the doubles in 1928. In 1931 RSM Marrison and SQMS P. S. Dundon won the other ranks doubles. SQMS H. F. Walker and SSgt J. Seymour had a share in the other ranks doubles title in 1927, 1929 and 1930. During the early 1930s, when Berger, Mahony, who retired in November 1929, and Fisher had all gone, there was a dearth of good young players. Major W. R. V. Warren and Captains I. K. Price, F. W. J. White, and G. Bright Holmes, who played for Devon while he was stationed at Devonport, kept the flag flying until the arrival of Lieutenant P. G. Turpin, a Somerset county player, Lieutenant S. H. A. Scroope, who came from a tennis playing family, and Lieutenant M. F. Turner-Cooke, all of whom played for the Army betweeen 1935 and 1939.

On the outbreak of war the well known lawn tennis referee, A. D. C. Macaulay (Lieutenant Colonel A. D. C. Macaulay OBE) joined the Royal Army Service Corps from the reserve of officers and served in the North African campaign with the Corps. After the war he became the Secretary of the All England Lawn Tennis Club, Wimbledon.

Tug-of-War and Boxing In December 1922 Feltham became the first entirely mechanised Royal Army Service Corps station. Here the Heavy Repair Shop (developed from a light repair unit in 1925), the Vehicle Reserve Depôt and the Mechanical Transport Stores Depôt, constituted the centre of workshops practice and training, and 'R' Depôt Company became the MT Driver Training School of the Corps.

Within the next few years Feltham won great destinction in two fields of sport: tug-of-war and boxing.

The tug-of-war team was first formed in 1928 (a team from ASC Aldershot had competed at Olympia before the war). In ten years, thanks largely to the remarkable expertise, amounting almost to genius, of Sergeant W. McCabe, who coached the team throughout the period, the Feltham teams triumphed over all Service and civilian opposition. The first real success came in 1933, when the team won the 130-stone Army championship at Olympia. Feltham repeated this success in 1934 and also won the 100-stone championship of the Amateur Athletic Association at the White City, which was conducted under slightly different rules. In 1935 Feltham were 130-stone Army champions for the third successive year and added the Inter-Services Championship to their list of successes, by beating the redoubtable Royal Marines team at Olympia, and once more won the AAA championship. Next year the Royal Marines had their revenge, but the Royal Army Service Corps retained the Army and AAA titles. In 1937 Sergeant McCabe became more ambitious and entered a team for the 110-stone championship. He was immediately successful. In addition to winning the 130-stone Army Championship for the fifth successive year, although the Inter-Services title went once more to the Royal Marines, Feltham won both the Army and Inter-Services 110-stone championship and the 100-stone AAA championship for the fourth consecutive year.

In 1938 came the grand slam. For the first time an Army team won all four titles – the Army and Inter-Services 110-stone and the Army and Inter-Services 130-stone championships. In addition the 100-stone AAA championship was won for the fifth time. And this record was repeated in 1939. Thus the Royal Army Service Corps, Feltham became 130-stone Army champions for the seventh year running and 110-stone champions for the third consecutive year, 130-stone Inter-Services champions for the second consecutive year (for the third time) and 110-stone Inter-Services champions for the third year in succession. It is doubtful whether such a record will ever be equalled. Sergeant McCabe's achievement was all the more remarkable, in that the members of his teams were constantly changing and every year the art of tug-of-war had to be taught to men who were new to the game. Among the names which appeared in the list of successful teams those of Bond (Major General G. A. Bond CB CBE) and Boxall (Major J. R. Boxall) appear on more than one occasion.

Boxing The Royal Army Service Corps has produced a number of good boxers. Before the 1914-18 war Sergeant J. Reynolds won the Army middle-weight championship in 1902. Lieutenant B. U. Farrell, a most promising boxer, won the Army light heavy-weight title and the Imperial Services Boxing Association championship in 1920 and, but for his untimely death in India, would no doubt have had a distinguished boxing career. Lieutenant A. B. Butterworth won the Army Officers title at featherweight and boxed for the Army against the Royal Navy in 1912. Sergeant Pape won the Army lightheavy-weight championship in 1921 and 1922. But by far the best Corps boxers were Lieutenant J. Courtis and Lieutenant H. C. D. Huxham. Courtis won a dozen Army and Imperial Services Boxing Association championships at middleweight and heavyweight. He was at his peak in 1924, when he was selected to represent Great Britain in the Olympic Games in Paris. Huxham transferred to the Royal Army Service Corps from the Loyal Regiment in 1932, with several Army lightweight championship wins already under his belt. When he retired from active boxing six years later in 1938 he had created an all time Army record by winning nine Army championships, at lightweight and welterweight, and six ISBA titles. Both he and Courtis played a great part in helping Feltham in its efforts to win the Army inter-unit boxing championship. A team from Feltham was first entered for the competition in 1931. In 1936 the Feltham team reached the final of the Inter-unit championship but lost to the Loyal Regiment. As runners-up they won the Noble Trophy. After losing in the final for two more years — by a single point in 1938 — the Feltham team at last achieved its ambition by winning the King's Trophy in 1939 by defeating the 1st Battalion the Welch Regiment by ten fights to five, and a score of 25 points to 20 at the Aldershot Command boxing centre. In presenting the trophy, the GOC-in-C Aldershot Command, Sir John Dill, congratulated the winners and commented that Feltham's triumph was the result of three main factors — hard training, the inspiration of Captain J. Courtis, and the full advantage which had been taken of competing with civilian teams. Since the team was first formed the training had been in the capable hands of Mr Matthews (Sergeant J. W. Matthews).

The Sassoon Cup had been presented by Sir Philip Sassoon, a former Under-Secretary of State for Air, for competition between the champion teams of the Army and the Royal Air Force. In 1938 the Feltham team was invited to deputise for the 2nd Battalion the King's Regiment, the Army champions, who had embarked for abroad before the date of the contest, and had only defeated Feltham by 23 points to 22. Feltham had a decisive victory over RAF Gosport by ten fights to one. In the following year, competing in their own right as holders of the King's Trophy, Feltham had to be content to share the cup with RAF Calshot, as the result of a tie.

Other individual achievements by Royal Army Service Corps boxers made an impressive record. Private Lacey from Feltham was Army and ISBA middleweight champion in 1932 and 1933; in 1934 he and Corporal Hughes boxed for the ISBA against the Police of Great Britain. Lance corporal G. Allday was awarded his Army boxing colours in 1936 and 1937. Lieutenant J. March was Army officers lightweight champion in 1937. Corporal Fisher (Feltham) won the Army flyweight title in 1938 and boxed for the Army; and in the year of Feltham's victory in 1939, Lance Corporal Colbourne (flyweight) and Lance Corporal Brown (middle-weight) were Army champions and Lance Corporal Brown also won the ISBA title, while Sergeant Fisher, Corporal Whitehouse and Lance Corporal Brown all represented the Army (three out of a team of eight).

Royal Army Service Corps boxing successes were not limited to Feltham. During the early 1930s the standard of boxing throughout the Corps steadily improved — no doubt, largely as a result of Feltham's efforts — and in 1933 the RASC Cairo team reached the Egypt Command final.

In Royal Army Service Corps sporting records the name of Feltham will always be associated with boxing and tug-of-war.

Athletics One of the outstanding athletes in the Royal Army Service Corps was G. M. Sproule (Major G. M. Sproule MC) who was Editor of the *Journal* from 1936 to 1939. His athletic career was cut short by the Great War. Between 1911 and 1914 he represented Oxford University in several events as a distance runner. He won the inter-varsity cross-country race in 1912 and 1913, and in 1914 he won the inter-varsity three mile race in the record time of 14 minutes 34⅕ seconds.

Another outstanding pre-war Royal Army Service Corps athlete was Major C. W. F. P. R. Dugmore. For many years he was the Army long and high jump Champion. He won the Midland Counties high and long jumps in 1908 and the Irish 120 yards hurdles in 1912. In the same year he represented Ireland against Scotland in the 120 yards hurdles. He represented Great Britain in the Olympic Games at the White City in 1908 in the long jump and the hop, step and jump. Private H. F. Thomas won both the mile and the three miles in the Army athletic championships in 1919, and he was followed by Corporal E. J. Slater, who won the mile in 1920 and 1921, and Sergeant A. W. Slade, who won the half-mile in 1920 and 1921 and the quarter-mile in 1921.

Apart from these successes and Feltham's achievements in the tug-of-war, the Royal Army Service Corps was not well represented in Army athletics between the wars. Second Lieutenant W. H. Summers won the 100 yards in the record time of 9.8 seconds in 1932 and 1933, but this was just before he transferred to the Royal Army Service Corps from the Duke of Wellington's Regiment.

In 1937 the first Royal Army Service Corps team was entered for the modern pentathlon. The team, which consisted of Lieutenant A. P. Campbell, Lieutenant J. March and Second Lieutenant R. Tadman, was placed fourth, with Lieutenant March third in the individual placings.

Swimming The Royal Army Service Corps registered a number of successes at swimming during the two decades under review. In 1925 the Corps won the swimming championships of the British Army of the Rhine, and in 1926 RASC Cairo District won the Egypt Command swimming championships and, ten years later, in 1936, won the command waterpolo league. In 1938, RASC Cairo had an even more successful season, winning the swimming championship cup and the knock-out waterpolo championship cup.

In the inter-Services swimming meeting between the Army and the Royal Air Force, seven Royal Army Service Corps swimmers were selected to represent the Army — MSS Major, L/Sergeant Gelsthorpe, Corporal Porter, Lance Corporal Hall, Drivers Ellis, Flower and Mannion.

At home, Service Companies Aldershot were successful in winning the inter-unit team swimming challenge shield in 1930 and, in the same year, won the Sir John French waterpolo cup. They repeated their win in the inter-unit championship in 1934 and in 1937 and 1938 won the unit team relay race (400 yards) in the Army swimming championships.

Outstanding individual performers were Private Rimmer, who won the 100 yards breast stroke in 1928 and 1929 in record time, and again in 1931; RSM A. J. Reid, who was the Army plunging champion in 1931 and 1932, Private Meeham who was spring board diving champion in 1923; Boy R. Mannion, who won the 100 yards (Boys) in 1932; and CSM Aldrich, who played waterpolo for the Army in the inter-Services championships. Sergeant Crookall and Lance Corporals Shuker and Ware also represented the Army, as did Drivers E. Wooller and Carruthers at diving and swimming.

Golf Many good golfers served in the Royal Army Service Corps. In the early 1920s the outstanding Corps player was Captain Smith (Major A. R. Smith MC). In 1925 he won the Black Watch Silver Medal, as runner up in the Army Championships and he continued to play good golf until he retired in 1939. In the same year (1925) Captain W. M. J. Hime was Army champion in Egypt and also won the Egyptian open championship. In 1923 Major J. C. L. Godfray, who was also a good cricketer, won the golf championship of Turkey, the first soldier to do so since the competition was instituted in 1905 – and in 1924 Lieutenant Colonel E. P. Blencoe DSO became the golf champion of Ceylon.

In 1934 the first of a batch of Cambridge University golfers joined the Royal Army Service Corps. Second Lieutenant M. R. Gardner, who had all the qualities of a first class amateur golfer, won the 'Bystander' £500 trophy for the second time in 1935 and followed this up by winning the Army golf championship (the Black Watch Gold Medal), the first Corps officer to do so. In 1936 he successfully defended his title as Army champion. The arrival of three more Cambridge players greatly strengthened the Royal Army Service Corps Training Centre team and, under the captaincy of that great games player, Captain G. B. McDonald, the Royal Army Service Corps won the Army inter-unit golf championship by a record margin. In addition to Captain McDonald the side included Lieutenants M. R. Gardner, W. F. Ellis, T. F. B. Law and P. J. Gold. Once before, in 1921, a Royal Army Service Corps team had been in the final, when it lost by one hole. The same team won the inter-unit championship again in 1938 and three of the team, Gardner, Ellis and Law, were selected to represent the Army against the Royal Navy and Royal Marines.

The individual Army championship was won again in 1939 by Lieutenant T. F. B. Law, who won the Black Watch Gold Medal at Sandwich with a score of 149, the first time that 150 had been broken in the championships.

Lieutenant Gardner was posted to Egypt early in 1939 and, within a few weeks, celebrated his arrival by winning the Egyptian open golf championship.

The high level performances of these few outstanding golfers should not be allowed to overshadow the very real ability of a number of other Royal Army Service Corps players who formed the backbone of the game in the Corps and successfully upheld its reputation between the two wars.

Mounted Sports Mention has already been made of the achievements of the Royal Army Service Corps coach. The first successes were largely due to the enthusiasm and energy of Major Pereira (Colonel A. B. P. Pereira DSO), who had the satisfaction of winning the coaching marathon at the Aldershot Show in 1921. Captain H. R. Kerr was fired with equal enthusiasm and during three years at Woolwich competed successfully in several coaching marathons. When Captain Kerr left Woolwich to take up the appointment of company commander at Sandhurst, Lieutenant Colonel E. R. O'Hara took charge of the coach and, in

1925, succeeded once more in winning the coaching marathon at the Aldershot Horse Show. Driver Freeman won the Silver Horn at the same time in the hornblowing competition.

When the coach moved to Aldershot with 'Y' Depôt Company in 1929, it came once more into the hands of Captain Kerr, now adjutant of the Training Centre, and 1930 was noteworthy in that the Corps coach took four first prizes, one third and two fourths in ten days, and then gained the first prize for military coaches in the Aldershot Command Show.

When Captain Kerr went to the Staff College, Lieutenant A. F. Somerset succeeded him as whip in 1931 and 1932. He had an auspicious start. In a single season the coach took first prize for regimental coaches in every event in which it competed. In 1933 and 1934 many of these successes were repeated.

Sadly, in 1935 the Royal Army Service Corps coaching activities came to a temporary end and the coach and equipment were bought by Major L. H. Burlton. The return of Lieutenant Colonel Kerr to Aldershot as chief instructor in 1937 led to a revival and, in 1938, the coach was bought back by the officers' mess and a programme of drives arranged during the next two years. The coach was destined to have a further lease of life and another run of successes in competitive classes after the Second World War.

Members of the Royal Army Service Corps were prominent at military race meetings and in the hunting field as long as horse transport units continued to exist, particularly at Woolwich, the traditional home of horse transport, at Bulford, and at overseas stations such as Malta, Gibraltar and Peshawar.

The Royal Army Service Corps point-to-point took place annually at Arborfield Cross, near the Remount Depôt, from its revival in 1921 until 1925, when it was cancelled because of insufficient entries. In 1927 the venue was shifted to Shipton Bellinger, near Bulford, where it flourished for five years, largely through the influence and enthusiasm of Colonel H. N. Foster CMG CBE and Lieutenant Colonel A. M. Wilson DSO who were two of the Corps's keenest horsemen. The point-to-point had always been dependent on the financial support of the Corps Club. When it was decided that Corps Club activities must be curtailed in 1932, owing to the recession, the point-to-point was one function to be cancelled. With no horse transport unit left in Bulford, it had become increasingly difficult to organise it, although for a time a combined event with the Royal Artillery was arranged, it was eventually decided that it must be abandoned.

Colonel O'Hara recorded a series of successful racing seasons in Malta. He was followed by Major C. Croker who, in addition to his racing successes, showed his prowess on the polo field and, in 1936, helped the Staff and Departmental team to win the King of Spain's polo cup.

Similarly, in Gibraltar, Major C. N. Custance and Lieutenant E. J. Pearn represented the Royal Army Service Corps in the Staff and Departmental team, which won the inter-regimental polo tournament in 1931. The former also won the Calpe Hunt Cup in 1933.

Captain J. R. Allen was the Corps's most accomplished performer in the show-jumping ring. Among his many successes, perhaps the most outstanding was at the International Horse Show in 1937 when, out of a total of 106 international entries, he shared second place on Hilton in the King George V Challenge Trophy.

Lieutenant A. F. Somerset won the first prize in the team jumping competition at the Aldershot Show and also came second in the international jumping contest at Olympia in 1930.

The success of the Animal (Transport) Company in winning the Sir Lindsay Cup for the first time in 1932 has been recorded elsewhere.

After the Second World War the Royal Army Service Corps continued to compete successfully in mounted events at the Royal Tournament at Earls Court and at the horse shows at Aldershot and Windsor.

Small-Arms Shooting Immediately after the First World War the best rifle shot in the Royal Army Service Corps was Lieutenant A. E. Martin, who was a leading member of the Army team. In 1921 the annual Royal Army Service Corps rifle meeting, which had been in abeyance since 1913, was revived, and the standard of shooting showed a marked improvement. In 1925 and 1926 Lieutenant H. C. Franklin was selected as a member of the Army revolver eight in the inter-Services matches against the Royal Navy and Royal Marines and the Royal Air Force.

In 1926 the Corps had two representatives in the Army Fifty (Captain H. P. Raymond and Private A. W. Bolton) and Sergeant Greenhead was in the Army Hundred. Captain Raymond had the reputation of being the best shot in the Aldershot Command. In the same year the Service Companies' team won the Aldershot Command small arms meeting light automatic challenge cup.

In 1931 Service Companies won three challenge cups at the Aldershot Command meeting, including the Chetwode Cup for anti-aircraft light automatics.

Of the Corps successes in revolver shooting, in 1937 the Training Centre team won the Duke of Connaught Cup and Trumpet Major Townsend became ARA revolver champion. The champion revolver team on that occasion included Major A. C. Booth MC, CQMS Gayton, Trumpet Major Townsend, Sergeant Burkimsher, Sergeant Firman and Sergeant Walton.

Fencing Captain Anderson (Colonel R. D. S. Anderson OBE), who transferred from the York and Lancaster Regiment in 1928, won Army and International Colours for fencing. He won the junior sabre championship of Great Britain in 1930 and was selected to represent the United Services of Great Britain in an international inter-Services assault-at-arms at Ostend. He also represented England against the United States and won his sabre match by five hits to three, and in 1931 he represented England against Scotland in the sabre event. Under his leadership the Royal Army Service Corps Training Centre team won the Aldershot Command inter-unit team fencing competition in 1931.

Cycling The Royal Army Service Corps has had one outstanding cyclist, Private H. A. G. Keates, who was a record-holder between 1931 and 1936 in a number of events. While serving in Northern Ireland and in Hong Kong his performances were almost unrivalled. Later, when the Army Cycling Union was formed after the Second World War, as Captain H. A. G. Keates, he became the honorary general secretary and leading member of the general committee.

Billiards Among indoor games, billiards has always had an enthusiastic following in the Corps. In 1914 Private Wright, of 53 Company, won the Army billiards championship, and in 1933 Lance Corporal Sweet narrowly lost the final of the Army championship.

Association football: No 1 Corps Depôt RASC, Aldershot, winners of the Army Cup 1922-23. The final between RASC v 2nd Bn Queen's Own Cameron Highlanders was played in the presence of The King & Queen, the Prince of Wales and Prince Henry; back row: Pte J. E. Clarkson (trainer); MSM A. Shelton; Pte H. Blondell; Cpl W. E. Wood; CSM E. Cooper; Cpl J. A. Quantrell (Reserve); centre: Lt F. W. Clarke; Lt W. S. Husher (Captain); Lt Col C. R. T. Annesly DSO; Capt E. H. Fitzherbert DSO MC; Lt K. E. Hegan; Lt G. McDonald MC; bottom: Pte H. Crossley; Sgt. R. D. McAvoy; Pte A. Kenning; Sgt E. G. Eastman. (Lt K. E. Hegan RASC was the most capped English Army player, having played for England on twenty occasions, and was Captain of the Army team from 1923 to 1929.)

ABOVE: RASC, Feltham Tug of War Team; 130 stone Army and Inter-Service Champions, 100 Stone AAA champions, 1935. BELOW: Association football: RASC Training Centre 1937 Winners of the Army cup. The final was played against the Training Battalion Royal Engineers; front: L to R: LCpl B. Watson; Pte J. W. Weir; Capt W. N. Craig-McFeeley; Lt A. P. Campbell; LCpl J. D. Gray; Cpl L. A. G. Sadler; rear: Pte A. G. Davies; Cpl P. W. Owen; Cpl A. G. Stewart; Dvr R. Campbell; LSgt G. V. Paul; LSgt F. C. Lambert; Cpl J. Dallas; SSgt J. Reidy; (inset) Cpl S. G. Waite.

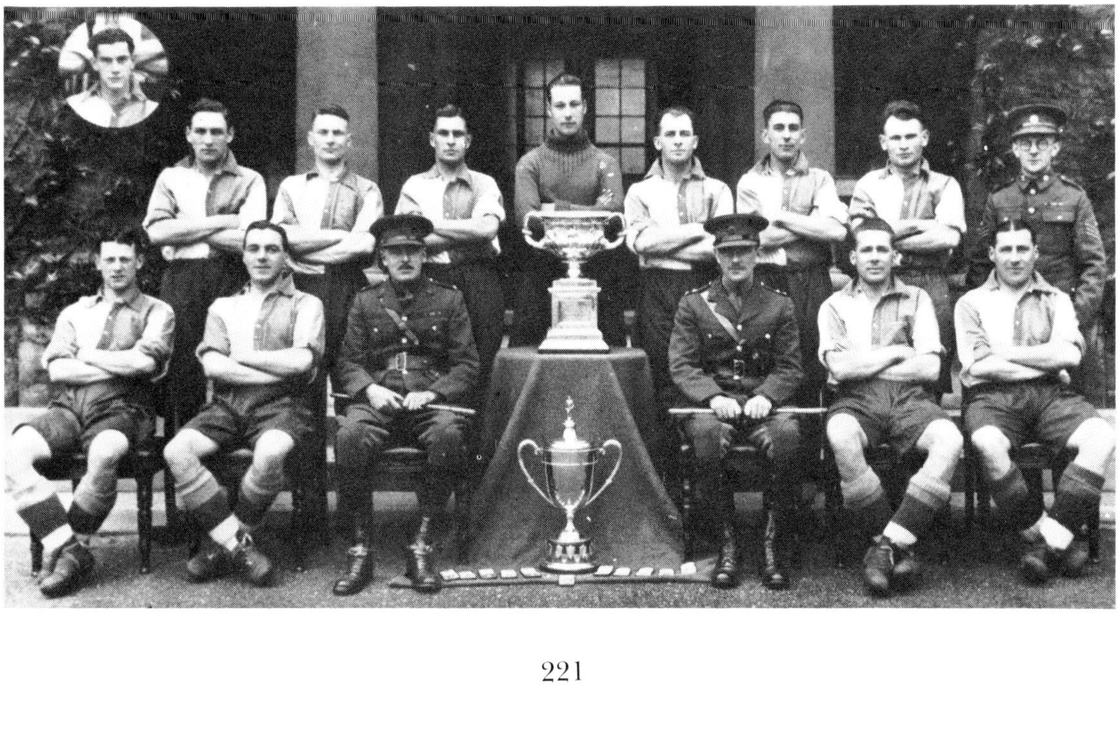

RUGBY FOOTBALL UNION.

TELEGRAPHIC ADDRESS:
"SCRUMMAGE, TWICKENHAM."

TWICKENHAM.

TELEPHONE:
RICHMOND 1708.

6 March 1922

DEAR SIR,

You have been selected to play for England v Scotland at Twickenham on Saturday 18th March 1922.

Please let me know by return if you can play. I will then send you all particulars.

Yours faithfully,

C. J. B. MARRIOTT,

RUGBY FOOTBALL UNION.

TELEGRAPHIC ADDRESS:
"SCRUMMAGE, TWICKENHAM."

TWICKENH
22nd March 1

Dear Sir,

I enclose you herewith your expenses for the Scottish Match. Please **return** the account receipted.

Will you order yourself an English Cap from Messrs George Lewin & Co Ltd. 8, Crooked Lane, Monument, E. C. 4, giving them your measurements.

Yours faithfully,

Charles J B Marriott
Secretary.

Lieut J. A. Middleton R.A.S.C.
Buller Barracks,
ALDERSHOT.

LEFT: Rugby Football: Lt J. A. Middleton RASC is selected for the English team, and RIGHT: instructed how to obtain his England Cap. BELOW: RASC Feltham Boxing team, Eastern Group and London District champions, later to win the King's Trophy in 1939.

RASC Cricket Team 1932; back: Mr Rowe; Sgt Varney; Dvr Hamlen; Capt F. W. Clark; LCpl Sadler; LCpl Stewart; Lt E. Dynes; front: Lt W. Gibb; Capt P. S. Whitcombe; Colonel W. N. White DSO, RSM F. Marrison, Capt T. E. Clarke MC. BELOW: RASC No 1 Unit winners Army inter-unit Golf Challenge Cup, 1937 & 1938; L to R: Lt P. J. Gold; Lt M. R. Gardner; Maj G. B. MacDonald MC; Lt N. F. Ellis; Lt T. F. B. Law. INSET: Lt J. Courtis, Olympic Bronze medallist, 1924.

Motorcycling — the start of the 1938 Army Cross-Country Championships.

Appendix A

SHORT BIOGRAPHIES

Field Marshal HRH Prince Arthur, the Duke of Connaught and Strathearn KG KT KP GCB GCSI GCMG GCIE GCVO GBE VD
Born 1 May 1850. Seventh child, third son of Queen Victoria.
Commissioned Royal Engineers after completing cadets' course at Royal Military Academy, Woolwich. Transferred to Royal Artillery, then the Rifle Brigade (for 5 years) and the 7th Hussars (for 4 years).
Commanded the Guards Brigade at Tel-el-Kebir, 1882.
Commanded a lst Class District in India and was appointed Commander-in-Chief of the old Bombay Army. Promoted Major General after 19 years' service. Returned home in 1893 to command Aldershot District. Retired from an active career in 1898.
Appointed Colonel, the Army Service Corps 2 Sep 1902.
Colonel, the Royal Army Service Corps 1918 - 1932.
Colonel-in-Chief, Royal Army Service Corps 3 Dec 1932 until his death on 16 Jan 1942.
Governor General and Commander-in-Chief of the Dominion of Canada 30 Jan 1911.
Honorary Colonel, the Ceylon Supply and Transport Corps (from Apr 1933 the Ceylon Army Service Corps) 3 Dec 1932.
As Patron attended the RASC Club Dinner 1921, 1925, 1927 and 1929.
Visited the newly-formed Militia Training Battalion, RASC 31 Aug 1939.
Died 16 January 1942.
General Rt Hon Sir Redvers Henry Buller VC GCB GCMG PC
Born 7 December 1839.
Gazetted Ensign in the 2nd Battalion The 60th Rifles (KRRC) 1858.
Served in China 1860. Present at the taking of the Taku Forts, and the capture of Pekin.
Red River Expedition 1870. Ashanti War 1874. Kaffir War 1878. Commanded the Frontier Light Horse. Zulu War 1878 - 79
(Brevet Lt Col, VC, CMG, Medal with clasp).
Chief of Staff, Boer War 1881.
Intelligence Department, Egypt 1882. Present at Kassassin and Tel-el-Kebir (despatches, KCMG, medal with clasp, 3rd Class Osmanieh, Khedive's star).
Soudan War 1884 (Despatches twice. Promoted Maj Gen. Two clasps). Chief of Staff Soudan 1884 - 85 (despatches, KCB, clasp).
Quartermaster General 1887.
Under-Secretary for Ireland 1887.
Adjutant General 1890 - 97. Promoted Lieut General 1891. General 1896.
In command at Aldershot 1898 - 99. (In succession to the Duke of Connaught). Served South Africa 1899 - 1900, first as General Commanding the Forces in South Africa, and afterwards as General Officer commanding in Natal. Conducted the operations for the relief of Ladysmith, which was successfully accomplished after an investment of 118 days (28 February 1901). He subsequently conducted operations resulting in the expulsion of the Boer Army from Natal (despatches). Commanded lst Army Corps, Aldershot 1901.
Retired 1906.
Died 2 June 1908.
General Buller was awarded the Victoria Cross for his bravery in the Zulu War of 1879. On 28 March during the retreat at Inhlobana, Brevet-Lieut Col Redvers Henry Buller, of the 60th Rifles, whilst hotly pursued by the Zulus, rescued Captain D'Arcy of the Frontier Light Horse and carried him on his own horse until he overtook the rearguard. He rescued two other men on the same day and he had repeatedly been brought to notice for gallant conduct.
Lieutenant General Sir Frederick Thomas Clayton KCB KCMG 'Father of the Corps'
Born 7 October 1855.
Commissioned Royal Warwickshire Regiment 29 Nov 1876.
Joined the Commissariat and Transport Staff 1882.
Appointed Deputy Assistant Commissary General (Captain) 1883.
Service in Bechuanaland and China 1883-1887.
Transferred to the Permanent List of the ASC 1 Apr 1889.
OC ASC Ashanti Expedition 1895. Promoted Brevet Lieut Col 25 Mar 1886.
Chief Instructor ASC School of Instruction 1896-1899.
Service in South Africa 1899-1900.
Recalled to command ASC, Aldershot 1900.
Appointed AQMG for ASC, War Office (virtual head of the Corps) 1902.
Assistant Director of Transport 1904-1905.

Director of Supplies, War Office (Brig General) 1906-1909.
Promoted Major General 2 Apr 1909.
Major General i/c Administration South Africa Command 1911.
Director of Supplies, BEF and Inspector of QMG Services in France 1914.
Inspector General of Communications (temp Lt Gen) 1915-1917.
Retired, with the Honorary rank of Lieut General 7 Oct 1917.
Colonel Commandant, Royal Army Service Corps Sep 1921-Oct 1925.
President RASC Club Nov 1928-Dec 1933. Vice Patron Regimental Association 1927.
Died 4 December 1933.
Decorations: CB 1901, KCMG 1915, KCB 1917. Cdr Order of Leopold (Belgium), Grand Officer, Legion of Honour (France), Grand Cordon of the Order of the Sacred Treasure (Japan), Croix de Guerre (Belgium).

Major General Nelson Graham Anderson CB CMG DSO

Born 14 September 1875.
Commissioned Royal Sussex Regiment 1896.
Service in South African War.
Seconded to Supply and Transport Corps, India 1903 - 1908.
Somaliland 1908 - 1910.
To Curragh as DAQMG 5 Division, 1913 and to France as AA & QMG 1914.
AQMG 7 Corps 1915.
DA & QMG 1 Corps (temporary Brigadier General) 1916.
DQMG Staff College 1919 - 1921.
AD ST HQ Eastern Command 1921 - 1924.
Promoted Major General 1 Jan 1924. ADC to the King 1922 - 1924. Retired Mar 1927.
Died 4 July 1945.

Colonel Hon Major General Sir Llewellyn William Atcherly Kt CMG CVO

Born 1 March 1871.
Commissioned East Lancashire Regiment 1890.
Transferred to ASC 1894.
Service Ashanti 1895 - 1896. South Africa 1899 - 1902. Brevet Major 1900. Great War 1914 - 1916.
Chief Constable Shropshire 1905 - 1908.
Chief Constable West Riding, Yorkshire 1908 - 1919.
HM Inspector of Constabulary 1919 - 1936. Re-employed as HMI 1940-1945.
Vice President of RASC Club Dec 1932.
Died 17 February 1954.

Major General Sir William George Bertram Boyce KCMG CB DSO

Born 27 May 1868.
Commissioned Royal Berkshire Regiment 1887.
Transferred to ASC 1898. On Staff Cork District 1899.
DAAG 2 Division South Africa 1899 - 1900.
AD ST Eastern Command 1913.
Director of Transport, British Armies in France Nov 1914 - June 1919.
Promoted Major General June 1917.
Retired 1920.
Appointed to the Board of Directors of Rolls Royce 1933.
Died 18 July 1937.
Order of St Stanislaus, 2nd Class with Swords. Order of Legion of Honour.

Major General Maxwell Spieker Brander CB OBE MIMechE

Born 11 October 1884.
Joined ASC from Royal Military College, Sandhurst 1906.
On completion of Junior Officers' Course attended a works course at John Fowlers' at Leeds. Instructor at MT School of Instruction, Aldershot.
Commanded 3rd Cavalry Ammunition Park 1914.
OC Base MT Depôt Rouen, and Calais 1914 - 1916.
Appointed to QMG3 branch, War Office 1917.
Instructor RASC Training Establishment 1922.
Chief Instructor (Brevet Lt Col) MT School of Instruction 1923 - 1926.
Service in India, HRS Bombay and Chaklala, 1916 - 1927.
Inspector and Chief Instructor of Tracked Vehicles under Director of Armaments, Woolwich Arsenal 1928 - 1932.
OCRASC Portsmouth 1933.
Commandant RASC Training Centre, Aldershot 1933 - 1936.
Promoted Major General Jan 1936.
Inspector RASC Aug 1936.
Director of Supplies and Transport 1937 - 1940.
Major General i/c Administration Eastern Command 1940 - 1941. Retired 1941.
Deputy Director General of Mechanisation (Wheeled Transport), Ministry of Supply 1941 - 1947.
Colonel Commandant RASC 1942 - 1949.
Died 30 October 1972.

Brigadier General Sir Charles Henry Bridge KCMG CB
Born 16 June 1852. Married Elizabeth Dorcas, daughter of Sir Edward Morris.
Entered the Control Department from Cambridge University 1872.
Service in the West Indies, Woolwich and Dublin.
Active service with the Commissariat and Transport Staff, Egyptian Campaign 1882.
Adjutant, Aldershot 1885.
DAQMG Army Headquarters 1886 - 1888. Worked with General Buller on conditions for the formation of the Army Service Corps. Promoted Lieut Colonel in ASC 1889. Commanded ASC, Aldershot 1891 - 1896.
Service South Africa. Matabeleland Campaign 1896. CB. Brevet Colonel.
Headquarters Staff, Colchester 1897 - 1899.
Director of Transport, South African Field Force 1899 - 1900. CMG.
AAG Aldershot 1900 - 1902. Director of Transport 1903. Inspector of Remounts 1903.
Retired 1903.
Re-employed as Assistant Inspector of Remounts. Organised the Remount Depôt at Arborfield Cross. Inspector Remount Depôt from 1906.
Promoted Brigadier General 1914.
Head of Mule Purchasing Commission in America 1914 - 1916. KCMG 1916.
Retired 1916.
Died 18 July 1926.
Major General Sir Herbert Napier Bunbury KCB
Born 15 February 1851.
Commissioned Royal Artillery 1871. Transferred to ASC 1889. The first officer to join the Corps from the Regimental List.
Deputy Assistant Commissary General Commissariat and Transport Staff 1880-1888. DAAG Headquarters ASC 1888 - 1890. DAAG Eastern District 1890 - 1891. DAQMG Army Headquarters 1891 - 1896. While in this appointment he signed the Order authorising the ASC Badge 1894.
Service in South African War: AAG for Supplies 1899. Assistant Director of Transport 1899 - 1900. Director of Transport (Colonel) 1900 - 1901. AAG Cork Distrrict 1901 - 1902. Director of Transport 3rd Army Corps 1902 - 1903.
Director of Supplies and Transport 3rd Army Corps 1903 - 1905.
AD of S and T Irish Command 1905.
Brigadier General i/c Administation Gibraltar 1906.
Major General i/c Administration Irish Command 1907 - 1908. (The first Major General in the Corps.)
Retired 1913. President Royal Army Service Corps Club 1919 - 1922.
Died 18 January 1922.
Major General Sir Evan Eyare Carter KCMG CB MVO
Born 11 August 1866.
Commissioned The Leicestershire Regiment 1889. Transferred to ASC 1890.
Staff College 1895 - 1896. One of the first Corps psc officers.
DAAG Belfast District and South Africa 1897.
South Africa War, staff of Lines of Communication and Army Headquarters.
DAQMG War Office 1902.
Chief Instructor ASC Training Establishment 1905 - 1908.
AD of ST London District 1913.
Colonel i/c Records 1914. AQMG Lines of Communication, France 1914.
Director of Supplies, British Armies in France, 1915 - 1919.
Promoted Major General 1918.
Director of Supplies and Transport, War Office 13 Mar, 1919 to 9 Aug 1921.
Retired 1921.
Colonel Commandant Royal Army Service Corps 1921 - 1933.
Hon Colonel 54th (East Anglia) Divisional Train RASC (TA).
Colonel-in-Chief First Aid Nursing Yeomanry.
Chairman of Regimental Association until his death.
Died 2 February 1933.
Commander Legion of Honour, Order of St Stanislaus 2nd Class.
Commandeur Ordre de la Couronne St Anne of Russia (2nd Class), Hon Associate St John of Jerusalem, Belgian Croix de Guerre.
US Distinguished Service Medal.
Major General Sir Alban Randall Crofton Atkins KCB CMG
Born 29 January 1870.
Commissioned Royal Marine Light Infantry 1889. Transferred ASC 1894.
Curragh 1894. Adjutant 1895.
Service Ashanti 1895 - 1896 and Sierra Leone 1899.
Service South African War 1899 - 1900. Brevet Major.
Adjutant ASC Aldershot 1901 - 1902.
Barbados 1902 - 1904.
Staff Captain and Deputy Assistant Director of Supplies, Army Headquarters 1904-08.

Chief Instructor ASC Training Establishment 1908 - 1911.
DAD of T War Office 1911 - 1914. Brevet Lt Col 1913.
AD of S and T London District 1914.
AD of T British Expeditionary Force 1914.
DD of S and T (temp Brigadier General) War Office 1915. Brevet Colonel. Succeeded Major General S. S. Long as Director of Supplies and Transport 11 Mar 1916.
Promoted Major General June 1918. Vacated appointment of D of S and T Jan 1919. Retired 13 April 1920. Died 29 April 1926.
Order of White Eagle 3rd Class (Serbia).

Major General George Freshfield Davies CB CMG CBE

Born 18 January 1872.
Commissioned Lincolnshire Regiment 1892. Transferred to ASC 1895.
Service Egypt 1898.
South Africa War 1899. Present at relief of Ladysmith.
Territorial Forces Adjutant, West Lancs Div Transport and Supply Column, Liverpool 1908.
DADST South Africa 1909.
DAA & QMG South Midland Division (TF) 1914.
DDST Mediterranean Expeditionary Force (temp Colonel) 1915.
DST Egyptian Expeditionary Force 1916 - 1919 (temp Brigadier General). DD of T War Office June 1919 - 1923.
Inspector Royal Army Service Corps (Colonel-on-the Staff) April 1923.
Director of Supplies and Transport War Office 9 Aug 1925 - 1929.
Retired Aug 1929.
Hon Colonel 43rd (Wessex) Divisional Train RASC TA until 1933.
Died 20 October 1936.
Orders of El Nahda and the Nile.

Major General Sir Reginald Ford KCMG CB DSO

Born 7 December 1868.
Commissioned Royal Marines 1889. Transferred to ASC 1893.
Commanded 6 Company ASC, Aldershot.
Assistant Instructor ASC Training Establishment, Aldershot 1899.
Service South Africa War (DSO). AAG.
ASC Training Establishment 1901.
OC ASC Singapore 1906 - 1908.
OC Service Companies, Aldershot.
AD of S, France 1914.
DAA & QMG 13 Corps 1915.
DQMG British Forces in France and Flanders 1918 - Nov 1919.
Promoted Major General June 1918. Retired 1920.
Emigrated to Rhodesia Feb 1920. On staff of Dunlops.
Colonel Commandant Royal Army Service Corps 1930 - 1938 (16 August).
Chief Traffic Commissioner, Southern Area 1931.
Chief Divisional Officer for London and the Home Counties for organising food control in the event of war.
Died 28 April 1951.

Major General Henry Nedham Foster CMG CBE AMIMechE

Born 19 Sep 1878.
Commissioned The Connaught Rangers. Transferred to ASC 1901.
Service South Africa War 1900 - 1902.
Chief Inspector of Subsidised Transport Vehicles 1911.
DAD of T War Office 1914.
AD of T (QMG 3) War Office 1915, Brevet Lt Col June 1916.
DD of T (Temp Colonel) 1917
DDST War Office 1919.
Chief Instructor, Aldershot Dec 1919 - Jan 1923.
Service in Malta.
Chief Inspector of Mechanical Transport 1924 - 1926.
ADST Southern Command 1927 - 1931. Promoted Major General 30 June 1931.
Retired from half-pay list 23 Aug 1933.
Re-employed HQ Eastern Command 1939 - 1941.
Wrote *History of Corps Point-to-Point.*
Died 1 October 1951.
Officer Legion of Honour.

Major General Sir Evan Gibb KBE CB CMG DSO

Born 12 March 1877.
Commissioned West India Regiment 1898. Transferred to ASC 1899.
Active Service on West Coast, Sierra Leone 1898.
Service South Africa War 1899 - 1902 (DSO). Relief of Ladysmith, Transvaal and Orange Free State.

Adjutant ASC, Aldershot 1905.
Staff Captain (1907) and Deputy Assistant Director (1910) QMG branch, War Office.
Instructor ASC Training Establishment, Aldershot Jan 1911.
DAQMG, BEF France 1914. AD of T Nov 1914. AQMG GHQ France 1915.
Director of Labour BEF (Brigadier General) 1916.
Controller of Salvage France 1918.
Brigadier General (DA & QMG) GHQ, British Troops in France and Flanders 1919.
General Officer Commanding the Forces in France and Flanders 1920.
AD of S and T. Brevet Colonel 1923.
AD of S and T, Aldershot Command (temp Colonel Commandant-on-the-Staff) Aug 1924.
Promoted Major General 15 June 1927.
Associated with NAAFI 1927 - 1929 (World tour).
Director of Supplies and Transport 1929 - 1933. Retired 1933.
Hon Colonel 47th (2nd London) Div RASC (TA) 1929 - 1934.
Hon Colonel First Aid Nursing Yeomanry Corps.
Colonel Commandant Royal Army Service Corps 1933 - 1947.
Joined firm of Sir Alexander Gibb & Partners (Consulting Engineers) 1934.
Director Herring Industry Board 1939 - 1940.
Director of Salvage 1940.
President London Chamber of Commerce 1943, 1944, 1945.
Died 13 July 1947.
Commander of the Crown (Belgium), Croix de Guerre (France), Croix de Boyaća (Colombia).

Major General Charles Arthur Hadfield

Born 27 March 1852.
Joined 2nd Derby Militia (Chatsworth Rifles) 1871. Commissioned North Staffordshire Regiment 1874.
Commissariat and Transport Staff as Deputy Assistant Commissary General 1881 - 1889.
Transferred to ASC 1889.
DAAG Portsmouth 1889 - 1892. OC ASC Curragh 1892 - 1895. DAAG HQ Ireland 1895 - 97.
OC ASC Dublin District 1898.
Director of Transport (Colonel) HQ 2nd Army Corps 1902 - 1903.
Director of Supplies and Transport HQ 2nd Army Corps 1903 - 1905.
AD of S and T Southern Command 1905.
AD of S and T Eastern Command 1906 - 1907.
Inspector of the Army Service Corps (Colonel) 1907 - 1909.
Promoted Major General 1908. The 3rd Officer to be promoted Major General in the Corps.
Major General i/c Administration South Africa Command 1909 - 1910.
Major General i/c Administration Ireland 1911 - 1912.
Retired Apr 1913.
Vice President Royal Army Service Corps Club 1912 and 1932.
Died 2 October 1938.

Lieutenant General Sir Antony Ernest Wentworth Harman KCB DSO

Universally known as 'Jacques'.
Born 1872.
Gazetted from Militia as 2/Lt in 3rd Dragoon Guards 1894.
Transferred to ASC 1896.
DADST in South Africa.
Rejoined 3rd Dragoon Guards as Adjutant 1908. Commanded 18th Hussars.
Service 1914 - 1918. Commanded Cavalry Brigade and Cavalry Division (DSO).
Commandant Cavalry School, Netheravon. Commanded lst Cavalry Brigade Aldershot 1920 -24.
Colonel-on-the-Staff i/c Administration Northern Command 1924 - 1926.
Commandant Equitation School and Inspector of Cavalry 1926 - 1930.
General Officer Commanding 1st Division, Aldershot 1930 - 1934. Retired Apr 1934.
ADC to HM the King 1923 - 1928.
Vice-President Royal Army Service Corps Club 1932. He continued to be a member until his death.
Died 26 September 1961.

Major General Percy Orr Hazelton CB CMG ADC

Born 2 March 1871.
Commissioned from Royal Military College Sandhurst.
Service South Africa War 1900 - 1902.
Chief Instructor ASC Training Establishment, Aldershot 1 Jan 1911 - 1 Jan 1914.
OCASC Shorncliffe 1914.
DAD of T and AD of T France 1914 - 1915.
D of S and T East Africa (temp Brigadier General) Jan 1916 - Mar 1918.
Assistant Director of Military Transport Woolwich 1918 - 1919.
DD S and T Rhine Army Apr to Sep 1919.
AD S and T Scottish Command 1919.

AD of T Home Forces 1919.
AD S and T Eastern Command 1920.
Director of Supplies and Transport War Office Aug 1921 - Aug 1925.
Promoted Major General 22 Dec 1921.
ADC to HM the King 1921 - 1922.
Died 14 December 1952.
Commander of the Order of the Crown of Belgium.

Major General Sir Charles Ernest Heath KCB CVO

Born 20 September 1854.
Commissioned 32nd Cornwall Light Infantry 1873. Adjutant 1877 - 1881.
Deputy Assistant Commissary General in Commissariat and Transport Corps 1881.
Transferred to ASC 1888. DAAG Malta Dec 1888 - 1892.
DAAG Southern District Headquarters 1892 - 1894.
OC ASC Dublin District 1894 - 1897.
Staff South Eastern District 1898 - 1901.
AQMG ASC Army Headquarters (War Office) 1901 - 1905.
Brigadier General i/c Administration Aldershot 1905 - 1907.
Major General i/c Administration Aldershot 1907. 2nd Corps Officer to be promoted Major General.
Awarded CVO after Royal Review by HM the King 12 June 1907.
Director of Transport and Remounts War Office 1907 - 1911.
Director of Quartering War Office 1914 - 1916.
Deputy Quartermaster General War Office 1916 - 1917 (KCB).
Retired 1917.
Vice President Royal Army Service Corps Club 1920.
Died 23 October 1936.

Major General Percy Eyre Francis Hobbs CB CMG

Born 18 February 1865.
Commissioned Royal Marine Light Infantry 1883. Transferred to ASC 1889.
DAAG 1899. Service South Africa War 1899 - 1900 (CMG).
Assistant Instructor and Chief Instructor ASC School 1901 - 1905. Editor of ASC *Journal* for four years. Brevet Colonel 1908.
Assistant Director of Supplies and Transport Irish Command and Eastern Command 1909 - 1913. Half pay 1913.
DA and QMG (temp. Brigadier General) Headquarters 1st Army 1914.
Promoted Major General 1915.
Retired 1917.
Colonel Commandant Royal Army Service Corps 1925 - 1935
First PMC of Buller Mess on its formation in 1895.
Died 26 October 1939.
Croix de Commandeur Légion d'Honneur (French).

Major General Frederick Francis Johnson CB OBE DL

Born 1 May 1852.
Commissioned 69th Foot 1874. 50th Regiment 1875. Attached C and T Staff 1881 - 89.
Deputy Assistant Commissary General Egyptian Campaign (Tel-el-Kebir) 1882.
DAAG and Chief Staff Officer Jamaica 1887 - 1890.
Transferred to ASC 1889. DAAG Belfast District 1891 - 1894.
OC ASC Woolwich 1899.
Assistant Director of Supplies South Africa war 1899 - 1900.
AAG North Eastern District 1900 - 1903.
ADST and AQMG Eastern Command 1903 - 1907. Retired 1907.
AQMG Southern Command Aug 1914.
Major General i/c Administration Southern Command 1915 - 1916. Honorary Major General June 1917.
Secretary Essex Territorial Forces Association 1917 - 1922.
Died 6 September 1931.

Major General Frederick William Brooke Koe CB CMG

Born 6 January 1862.
Commissioned Royal Marine Light Infantry 1881. Transferred to ASC 1889.
Service West Coast of Africa 1893 - 1894.
OC ASC China 1900. Brevet Lt Colonel.
OC ASC and ADST Pretoria 1907 - 1911. Half pay 1911.
AQMG Western Command 1912.
Assistant Director of Transport 1914.
Director of Supplies and Transport Mediterranean Expeditionary Force, Gallipoli.
(Brigadier General) 1915 - 1916. Promoted Major General 1916.
Inspector-General of Communications MEF Aug 1916 - March 1917.
Commissioner of Salvage East Africa 1918 - 1919. Retired 1920.

Vice President Royal Army Service Corps Club 1920.
Master of a pack of harriers at Nenagh, Co Tipperary 1926 - 1933.
Died 6 February 1935.

Major General Sir Frederick William Bainbridge Landon KCMG CB

Born 27 February 1860.
Commissioned 76th Regiment 1879. Transferred to ASC 1889.
DAAG Jamaica 1890 - 1895.
Second in Command and Supply and Transport Officer Benin Expedition 1897. Bt Lt Col.
South Africa War, DAAG Natal 1899. Assistant Director and Director of Supplies 1900 - 1902. Brevet Colonel 1900.
OC ASC Aldershot 1902. OIC Records (Colonel) and Depôt Commander Woolwich Dockyard 1905.
Assistant Director of Supplies and Transport Aldershot Command 1908 - 1909.
Inspector Army Service Corps (Brigadier General) Headquarters of the Army 1909 - 1912.
Director of Transport and Movements War Office (Major General) 1913 - 1914.
Inspector of the Army Service Corps and QMG's Services 1914.
Chief Inspector of QMG Services 1918.
Retired 1st October 1919.
Colonel Commandant Royal Army Service Corps 1921 - 1930.
Hon Colonel East Anglia Transport and Supply Column TF 1914 - 1922.
Hon Colonel 44 (Home Counties) Div, RASC TA 1925 - 1930.
Vice President Royal Army Service Corps Club 1920.
President Royal Army Service Corps Club Dec 1933.
Vice Patron RASC Regimental Association 1934.
Died 26 October 1937.

Major General Sidney Seldon Long CB

Born 31 March 1863.
Son of Commissary-General J. Long JP.
Commissioned Durham Light Infantry from Militia 1884. Transferred to ASC 1889.
DAAG South China 1896 - 1899.
South Africa War 1899 - 1902. DAAG for Transport. Later AAG. Brevet Lt Col.
Commandant ASC Training Establishment, Aldershot 1908 - 1909.
OC Supply Reserve Depot Woolwich 1909.
Assistant Director of Supplies 1909 - 1912.
Director of Supplies and Quartering (Brigadier General) 1913 - 1914.
Director of Supplies and Transport, War Office 1914 - 1916.
Retired 11 March 1919.
Managing Director of Lever Bros Ltd.
President Royal Army Service Corps Club Nov 1937.
Vice Patron RASC Regimental Association 1938.
Died 31 January 1940.

Major General Charles William Macleod CB CMG DSO

Born 25 May 1881.
Joined ASC 1899.
Service South Africa War 1900 - 1902. Orange River Colony Operations 1900.
OC 20 Company ASC Aldershot.
Deputy Director of Transport GHQ France 1914 - 1918.
AD of T War Office 1919 - 1923. Brevet Colonel.
Assistant Director of Supplies and Transport Eastern Command 1924.
ADC to HM the King 1928. Half pay 1928.
Inspector of the Royal Army Service Corps 1929 - 1933.
Director of Supplies and Transport War Office 9 Aug 1933 - 14 Jun 1937.
Promoted Major General 1934. Retired 14 June 1937.
Colonel Commandant Royal Army Service Corps 1938 - 1944.
Staff of Director General of Munitions Production, Area Progress Officer for Scotland 1937.
Deputy Director of Supplies and Transport (re-employed) 1939. Reverted to retired pay 1944.
Died 18 October 1944.

Commissary General Sir Edward Morris KCB

Born 1833.
Entered Commissariat Department 1854.
Service in Crimea 1854 - 1856. Zulu War 1879. Egyptian Campaign 1882.
Egyptian Star and 2nd Class of the Medjidie. KCB 1882.
Commissary General 1881 - 1886. The last Commissary General.
Retired 1886. Adviser to General Buller on the formation of the ASC.
JP City and County of Dublin.
Died 15 February 1923.

Major General Harry Neptune Sargent CB CBE DSO
Born 6 April 1866.
Commissioned 2 Devon Regiment from Militia 1886. Transferred to ASC 1890.
Adjutant 1895 - 1897. Nile Expedition 1898.
South Africa War 1899 - 1902. DAAG. Present at relief of Ladysmith. (DSO).
Lieutenant Colonel 1906. Colonel 1909.
OIC Army Service Corps Records 1912.
AD of S and T 1913.AQMG 1st Corps 1914.
DA & QMG (Brigadier General) 1st Corps 1914 - 1916.
DA & QMG Reserve and Fifth Army 1916 - 1917.
Chief of British Mission, HQ Services of Supply, American Expeditionary Force, 1918 to July 1919.
Retired with the Honorary rank of Major General 16 Oct 1919.
Died 6 February 1946.

Major General Richard Talbot Snowden-Smith CB CBE MIMechE
Born 23 April 1887.
Commissioned ASC from Royal Military College, Sandhurst 1906.
First ASC Officer to fly an aeroplane 1910. Holder of 29th Aviation Certificate issued by the Royal Aero Club. Took part in aeroplane race from Brooklands to Brighton and return, organised by the Royal Aero Club and the Fédération Aéronautique Internationale, May 1911. Came second in a Farman biplane.
Staff Captain War Office 1915. AD of S and T 1917. DD of S and T 1919.
Inspector of Subsidised Mechanical Transport 1919. Chief Inspector (MT) RASC Training College, Aldershot 1926 - 1930.
Brevet Lieutenant Colonel 1927.
Deputy Assistant Director of Supplies War Office 1931 - 1933.
Commanded RASC Depot Feltham 1933 - 1934.
Assistant Director of Supplies and Transport QMG 6, War Office 1934 - 1937.
Inspector Royal Army Service Corps 1937 - 1940. Promoted Major General 2 Feb 1938.
Director of Supplies and Transport War Office 1940 - 1943. Retired 1943.
President RASC Regimental Association 1943 - 1946.
Died 14 August 1951.

Major General William Knapp Tarver CB CMG
Born 2 November 1872.
Commissioned Cheshire Regiment 1892. Transferred to ASC 1896.
Service in Jamaica and Ireland.
Instructor ASC Training Establishment, Aldershot.
Deputy Assistant Director of Supplies War Office Nov 1911.
Assistant Director of Supplies QMG 6 War Office 1914.
Service in Macedonia, Serbia, Bulgaria and Dardanelles.
Deputy Director of Supplies War Office 1917. Brevet Colonel 1917.
Deputy Director of Supplies and Transport (Brigadier General).
Temporary appointment as Director of Supplies and Transport War Office 20 Jan 1919 to 13 Mar 1919 (Temp Major General).
DD of S and T 1919. Assistant Director of Supplies and Transport, Western Command 1920. Promoted Colonel Aug 1920.
ADST QMG 6 War Office 1923 - 1925.
Inspector Royal Army Service Corps 1925 - 1929. Promoted Major General June 1925.
Chairman RASC Regimental Association 1934.
Colonel Commandant Royal Army Service Corps 1935 - 1942.
Died 8 April 1952.

Colonel Sir Edward Willis Duncan Ward Bart GBE KCB KCVO
Born 17 December 1853.
Sub-assistant Commissary in the Control Department 1874.
Assistant Commissary in the Commissariat and Transport Department 1876.
Adjutant Woolwich 1882 - 1885.
Sudan Expedition 1885.
Promoted Major in the ASC (Supernumerary List from the Commissariat and Transport Staff) 1889. Lieutenant Colonel 1890.
DAAG HQ Ireland 1892. AAG Ashanti Expedition 1895 - 1896.(CB).
DAAG Home District 1895 - 1899. Hon Secretary of the Royal Military Tournament 1895 - 1899.
Brevet Colonel 1898.
AAG South Africa 1899 - 1900. At relief of Ladysmith (KCB).
Director of Supplies South African Field Force 1900. Appointed by Lord Roberts to take over all the supply arrangements at Army Headquarters. AQMG 1901.
Retired from the Army 1901.
Appointed Permanent Under-Secretary of State at the War Office 1901 - 1914 (Jan).
Appointed Secretary of the War Office (the first Secretary of the Army Council) February 1904.
Honorary Colonel the Canadian Army Service Corps 15 April 1904.

Honorary Colonel 2nd London Divisional Transport and Supply Column ASC (TF) 1908.
Chairman of the County of London Territorial Force Association 1908.
Retired Jan 1914 (Baronetcy).
Commanding and Chief Staff Officer of the Metropolitan Special Constabulary 1914 to November 1918.
Director General of Voluntary Organisations.
Representative of the Secretary of State for War on all questions relating to the welfare and comfort of Colonial Troops.
Chairman Royal Army Service Corps Club 1919 - 1928.
President Union Jack Club.
Died 11 September 1928.
Officer of the Legion of Honour. Knight of Grace of St John.

Major General Frederick Maurice Wilson CB CMG

Born 19 November 1868.
Joined Army 1890. Royal Lancaster Regiment. Ashanti expedition 1895 - 1896. Transferred to ASC 1897.
Commandant ASC Training Establishment 1913 - 1914.
Assistant Director of Supplies, Deputy Director of Supplies, Deputy Director of Supplies and Transport (temp Brigadier General 1917) 1914 - 1918.
Director of Supplies and Transport Rhine Army Nov 1918 - Nov 1919.
Assistant Director of Supplies and Transport Eastern Command 1920.
Director of Supplies and Transport India 17 Aug 1921 - 1925.
Retired 15 September 1925. Colonel Commandant Indian Army Service Corps 1930.
Died 16 August 1956.

Appendix B

SENIOR APPOINTMENTS 1919-1939

Colonel of the Army Service Corps — 1902-1918
Colonel of the Royal Army Service Corps — 1918-1932
Colonel-in-Chief of the Royal Army Service Corps — 1932-1942

Field Marshal HRH the Duke of Connaught and Strathearn KG KT KP GCB GCSI GCMG GCI GCVO GBE VD personal ADC to the King.

Colonels Commandant

The appointment of three Colonels Commandant to the Royal Army Service Corps was approved by His Majesty the King in August 1921. The first appointments were announced in the *London Gazette* of 23 September 1921. The tenure of the appointment was for ten years or to the age of 70. The number was increased to four in 1950 and to five in 1970.

1922-1925 Lt Gen Sir Frederick Clayton KCB KCMG.
1922-1930 Maj Gen Sir Frederick Landon KCMG CB.
1922-1933 Maj Gen Sir Evan Carter KCMG CB MVO.
1925-1935 Maj Gen P. E. F. Hobbs CB CMG.
1930-1938 Maj Gen Sir Reginald Ford KCMG CB DSO.
1933-1947 Maj Gen Sir Evan Gibb KBE CB CMG DSO.
1935-1942 Maj Gen W. K. Tarver CB CMG.
1938-1944 Maj Gen C. W. Macleod CB CMG DSO.
1942-1949 Maj Gen M. S. Brander CB OBE.
1944-1954 Lt Gen Sir Humfrey Gale KBE CB CVO MC.

Directors of Supplies and Transport

Before 1913 there was a Director of Movements and Quartering, who was responsible to the Quartermaster General. In 1913 the appointment was split and Major General F. W. B. Landon was appointed Director of Transport and Movements and Brigadier General S. S. Long became Director of Supplies and Quartering. In the following year another change was made and Brigadier General Long was appointed the first Director of Supplies and Transport in 1914 and Major General C. E. Heath became Director of Quartering. A separate Director of Movements was appointed in 1915. Major General Long relinquished the appointment of Director of Supplies and Transport in 1916.

1914 Maj Gen S. S. Long CB.
1916 Maj Gen Sir Alban Crofton Atkins KCB CMG.
1919 Maj Gen W. K. Tarver CB CMG (Acting).
1919 Maj Gen Sir Evan Carter KCMG CB MVO.
1921 Maj Gen P. O. Hazelton CB CMG.
1925 Maj Gen G. F. Davies CB CMG CBE.
1929 Maj Gen Sir Evan Gibb KBE CB CMG DSO.
1933 Maj Gen C. W. Macleod CB CMG DSO.
1937 Maj Gen M. S. Brander CB OBE MIMechE.
1940 Maj Gen R. T. Snowden-Smith CB CBE MIMechE.
1943 Maj Gen Sir Reginald Kerr KBE CB MC.

Inspectors of the Army Service Corps and Royal Army Service Corps (Inspectors of Royal Army Service Corps Services)

The post of Inspector of the Army Service Corps was first authorised on 16 January 1907, when Colonel C. A. Hadfield was appointed. On his promotion to Major General two years later, he was succeeded by Colonel F. W. B. Landon in January 1909, in the rank of Brigadier General. He held the appointment until the end of 1912. In January 1913 Brigadier General G. R. C. Paul was appointed and he was succeeded in the following September by Colonel T. D. Foster, who was promoted to Brigadier General.

In 1914 the appointment was redesignated 'Inspector of the Army Service Corps and QMG's Services' and Major General F. W. B. Landon was re-appointed and remained in the post until his retirement on 1st October 1919. In 1918 the title became 'Chief Inspector of QMG Services'.

There was no Inspector of the Army Service Corps between the date of General Landon's retirement and the appointment of Colonel G. F. Davies as Inspector on 1 April 1923, in the new rank of Colonel-on-the Staff.

Inspectors of the Army Service Corps and of the Royal Army Service Corps

1907-1909	Colonel C. A. Hadfield.
1909-1912	Brig Gen F. W. B. Landon CB.
1913	Brig Gen G. R. C. Paul CMG ADC.
1913	Brig Gen T. D. Foster CB MVO.
1914-1919	Maj Gen F. W. B. Landon CB. (Sir Frederick Landon KCMG CB).
1923-1925	Colonel-on-the-Staff G. F. Davies CB CMG CBE.
1925-1929	Maj Gen W. K. Tarver CB CMG.
1929-1933	Brigadier C. W. Macleod CMG DSO ADC.
1933-1936	Brigadier W. N. White CB DSO.
1936-1937	Maj Gen M. S. Brander OBE MIMechE.
1937-1940	Brigadier (Maj Gen) R. T. Snowden-Smith CBE MIMechE.
1940-1943	Maj Gen E. H. Fitzherbert CBE DSO MC.
1943-1946	Maj Gen T. W. Richardson OBE.

General Officers of the Royal Army Service Corps other than Directors and Inspectors

Maj Gen N. G. Anderson CB CMG DSO, promoted Major General 1 January 1924.
Maj Gen Sir William G. Boyce KCMG CB DSO, Director of Transport, British Armies in France 1914-1919. Retired 1920.
Maj Gen Sir Reginald Ford KCMG CB DSO, DQMG British Forces in France and Flanders 1918-1919. Retired 1920.
Maj Gen H. N. Foster CMG CBE AMIMechE promoted Major General 30 June 1931. Retired 23 August 1933, from Half Pay List.
Maj Gen P. E. F. Hobbs CB CMG: Retired 1917.
Maj Gen H. N. Sargent CB CBE DSO. Chief of British Mission, HQ Services of Supply, American E F 1918-1919.
Maj Gen F. M. Wilson CB CMG. Director of Supplies and Transport, India 1921-1925. Retired 15 September 1925.

Appendix C

RASC UNIT LOCATIONS — NOVEMBER 1924

(excluding units in India)

Horse Transport Companies

'Y' Depôt Company	Woolwich	4 Company	Rhine
1 Company	Aldershot	5 Company	Rhine
2 Company	Bulford	24 Company	Aldershot
3 Company	Woolwich	25 Company	Egypt

Mechanical Transport Companies

6 Company	Aldershot	20 Company	London
7 Company	Aldershot	21 Company	Colchester
8 Company	Aldershot	22 Company	Shorncliffe
9 Company	Aldershot	23 Company	Devonport
10 Company	Bulford	26 Company	Belfast
11 Company	Bulford	27 Company	Londonderry
12 Company	Woolwich	28 Company	Rhine
13 Company	Portsmouth	29 Company	Rhine
14 Company	Bovington	30 Company	Egypt
15 Company	York	31 Company	Egypt
16 Company	Preston	32 Company	Malta
17 Company	Woolwich	'P' Company	Aldershot
18 Company	Chatham	'Q' Depôt Company	Aldershot
19 Company	Catterick	'R' Depôt Company	Feltham

Supply Companies

'A' Depôt Company	Aldershot	'G' Company	Bulford
'B' Company	London	'H' Company	York
'C' Company	Aldershot	'I' Company	Chester
'F' Company	Woolwich	'J' Company	Leith

Appendix D

STATIONS IN INDIA

where RASC Transport Units were located 1922-1928

1 (Indian) MT Company	Lahore
2 (Indian) MT Company	Wazir Force
3 (Indian) MT Company	Bombay
4(Indian) MT Company	Karachi
5(Indian) MT Company	Dera Ismail Khan
6 MT Company (2 MAC)	Rawalpindi
7 (Indian) MT Company	Lahore
8 (Indian) MT Company	Manzai
9(Indian) MT Company	Bangalore
10 (Indian) MT Company	Bareilly
11 (Indian) MT Company	Bareilly
12 (Indian) MT Company	Bombay
13 (Indian) MT Company	Bannu
14 MT Company (3 MAC)	Manzai
15 (Indian) MT Company	Wazir Force
16 (Indian) MT Company	Quetta
17 Mobile Repair Unit	Peshawar
18 Mobile Repair Unit	Dera Ismail Khan
19 Mobile Repair Unit	Quetta
20 Mobile Repair Unit	Rawalpindi
21 (Indian) MT Company	Wazir Force
23 (Indian) MT Company	Calcutta
24 (Indian) MT Company	Wazir Force
25 (Indian) MT Company	Wazir Force
26 MT Company (4 MAC)	Peshawar
27 MT Company (5 MAC)	Dera Ismail Khan
28 MT Company (6 MAC)	Quetta
31 MT Company (9 MAC)	Bangalore
32 MT Company (10 MAC)	Bareilly
33 (Indian) MT Company	Roorkee
34 (Indian) MT Company	Peshawar
35 (Indian) MT Company	Ferozepore
36 (Indian) MT Company	Roorkee
'A' MT Column	Peshawar
'B' MT Column	Dera Ismail Khan
'C' MT Column	Quetta
'D' MT Column	Chaklala
'E' MT Column	Poona
Base MT Workshops	Chaklala
Central MT Stores Depôt	Chaklala
Reserve Vehicle Park	Chaklala
1st Advanced Workshops	Peshawar
Technical Inspector Northern Circle	Peshawar
Technical Inspector Central Circle	Rawalpindi
Technical Inspector Southern Circle	Poona
MT Training School	Sitapore
Khyber Ropeway Company	Jamrud

Appendix E

RASC STATIONS — 1937

HOME

Aldershot Command

Aldershot

Training Battalion	7 Company	24 Company	40 Company
'A' Company	8 Company	41 Company (Bordon)	42 Company
'C' Company	9 Company	29 Company	
6 Company	Detachments at Blackdown	35 Company	

Eastern Command

Deptford		*Chatham*	
RASC Record Office	Supply Reserve Depôt	18 Company	
Kensington Barracks		*Colchester*	
'B' Company	20 Company	21 Company	
43 Company	44 Company	*Shorncliffe*	
Woolwich		22 Company	46 Company
'F' Company	17 Company	*Feltham*	
37 Company	38 Company	Vehicle Reserve Depôt	RASC Driving School
45 Company		MT Stores Depôt	MT Heavy Repair Shop

Detachments at: Brighton, Hertford, Hounslow, Norwich, Sheerness, Shoeburyness, Warley, Weedon, Windsor.

Northern Command

Catterick		*York*
19 Company	28 Company	15 Company
34 Company	51 Company	Detachments at: Derby, Newcastle, Lichfield.
'H' Company		

Southern Command

Bulford		*Portsmouth*	
'G' Company	10 Company	13 Company	49 Company
27 Company	36 Company	*Devonport*	
47 Company	48 Company	23 Company	
Larkhill			
11 Company			

Detachments at: Gosport, Parkhurst (I of W), Netley, Tidworth, Salisbury, Southampton, Oxford and Bovington.

Western Command

Chester	*Preston*	Detachments at:
'I' Company	16 Company	Pembroke Dock, Burscough

Scottish Command

Edinburgh

50 Company — 'J' Company

Detachments at: Glasgow, Perth, Inverness, Stirling.

Northern Ireland District

Belfast		*Londonderry*
26 Company	53 Company	54 Company

Detachment at Ballykinlar.

South Irish Coast Defences

Detachments at: Spike Island, Bere Island.

ABROAD

China Command

Hong Kong Area	*Tientsin Area*	*Shanghai Area*
12 Company	Detachment	Detachment

Egypt Command

Cairo		
D Company	31 Company	30 Company
25 Company	5 Company	39 Company
Palestine		
14 Company		

Detachments at: Cyprus, Alexandria, Port Said, Moascar, Khartoum.

Other Stations

Ceylon			
Detachment			
Gibraltar	*Jamaica*	*Malta*	
33 (Local) Company	Detachment	32 Company	
Mauritius	*Malaya* (Singapore)	*Bermuda*	*Sierra Leone*
Detachment	52 Company	Detachment	Detachment

ALLIED ARMY SERVICE CORPS

The Army Service Corps of the Dominion of Canada.
The Army Service Corps of the Commonwealth of Australia.
The Army Service Corps of the Dominion of New Zealand.
The Service Corps of the Union of South Africa.
Ceylon Army Service Corps.

TERRITORIAL ARMY

Eastern Command

54th (The East Anglian) Divl RASC. Chelmsford.
44th (The Home Counties) Divl RASC. Croydon.
1st Anti-Aircraft Divl RASC. Plumstead, SE 18.
The London Divl RASC, Chelsea SW 3.

Northern Command

50th (The Northumbrian) Divl RASC. Hull.
49th (The West Riding) Divl RASC. Leeds.
46th (The North Midlands) Divl RASC. West Bromwich.

Scottish Command

51st (The Highland) Divl RASC. Perth.
52nd (The Lowland) Divl RASC, Glasgow.

Southern Command

48th (The South Midland) Divl RASC. Birmingham.
43rd (The Wessex) Divl RASC. Plymouth.

Western Command

2nd Cavalry Divl RASC. Birkenhead.
53rd (The Welsh) Divl RASC. Swansea.
55th (The West Lancashire) Divl RASC. Southport.
42nd (The East Lancashire) Divl RASC. Manchester.

Appendix F

LOCATIONS OF RASC COMPANIES

The following tables show the location of the sixty-eight regular Transport Companies and the nine Supply Personnel Companies at the beginning of 1939.

Command or District	*Companies*	
	Transport	*Supply Personnel*
Aldershot	13	2
Eastern	12	1
London	3	1
Northern	6	1
Northern Ireland	3	0
Scottish	1	1
Southern	13	1
Western	2	1
China	1	0
Egypt	8	1
Gibraltar	1 (local)	0
Malta	1	0
Palestine	3	0
Singapore	1	0
	68	9

Of the sixty-eight Transport Companies thirty seven existed when the Army Service Corps was formed in December 1888. The following table shows the location of these Companies in December 1888 and fifty years later in December 1938.

December 1888	*December 1938*
1 Company Aldershot (Transport Depôt)	Aldershot
2 Company Woolwich (Transport Depôt)	Woolwich
3 Company Chelsea	Aldershot
4 Company Shorncliffe	Bulford
5 Company Aldershot	Egypt
6 Company Woolwich	Aldershot
7 Company Dublin	Aldershot
8 Company Curragh	Aldershot
9 Company Egypt	Bordon
10 Company Aldershot	Bulford
11 Company Devonport	Larkhill
12 Company Dublin	Hong Kong
13 Company Woolwich	Portsmouth
14 Company Woolwich	Palestine
15 Company Dublin	York
16 Company Aldershot	Preston
17 Company Aldershot	Palestine
18 Company Aldershot (Supply Depôt)	Chatham
19 Company Chatham	Catterick Camp
20 Company Aldershot	Kensington Barracks
21 Company Portsmouth	Colchester
22 Company Aldershot	Shorncliffe
23 Company Aldershot	Devonport
24 Company Shorncliffe	Aldershot
25 Company Chatham	Egypt
26 Company Woolwich	Belfast
27 Company Curragh	Bulford
28 Company Dublin	Catterick Camp
29 Company Chelsea	Aldershot
30 Company Aldershot	Egypt
31 Company Portsmouth	Egypt
32 Company Curragh	Malta
33 Company Woolwich	Gibraltar (local company)

December 1988	*December 1938*
34 Company Woolwich	Catterick Camp
35 Company Curragh	Aldershot
36 Company Aldershot	Bulford
37 Company Devonport	Woolwich
38 Company –	Woolwich
39 Company –	Egypt
40 Company –	Aldershot
41 Company –	Aldershot
42 Company –	Aldershot
43 Company –	London
44 Company –	London
45 Company –	Woolwich
46 Company –	Shorncliffe
47 Company –	Warminster
48 Company –	Warminster
49 Company –	Portsmouth
50 Company –	Edinburgh
51 Company –	Catterick Camp
52 Company –	Malaya
53 Company –	Northern Ireland
54 Company –	Londonderry
55 Company –	Bulford
56 Company –	Bulford
57 Company –	Colchester
58 Company –	Palestine
59 Company –	Colchester
60 Company –	Woolwich
61 Company –	Palestine
62 Company –	Bulford
63 Company –	Bulford (1 Apr 1939)
64 Company –	Aldershot
65 Company –	Egypt
66 Company –	Egypt
67 Company –	Palestine
68 Company –	Palestine

Appendix G

DIVISIONAL RASC (TERRITORIAL ARMY) 1939

Mobile Divisional RASC (TA)
42nd (The East Lancashire) Divisional RASC (TA) (501, 502, 503 Companies)
43rd (The Wessex) Divisional RASC (TA) (504, 505, 506, 547 (non-div) Companies)
44th (The Home Counties) Divisional RASC (TA) (507, 508, 509, 548 (non-div) Companies).
The London Divisional RASC (TA) (513, 514, 515, 549 (non-div) Companies)
48th (The South Midland) Divisional RASC (TA) (516, 517, 518, 550 (non-div) Companies)
49th (The West Riding) Divisional RASC (TA) (519, 520, 521 Companies)
50th (The Northumbrian) Divisional RASC (TA) (522, 523, 524, 552 (non-div) Companies)
51st (The Highland) Divisional RASC (TA) (525, 526, 527, 553 (non-div) Companies)
52nd (The Lowland) Divisional RASC (TA) (528, 529, 530, 554, (non-div) Companies).
53rd (The Welsh) Divisional RASC (TA) (531, 532, 533, 555 (non-div) Companies)
54th (The East Anglian) Divisional RASC (TA) (534, 535, 536, 556 (non-div) Companies)
55th (The West Lancashire) Divisional RASC (TA) (537, 538, 539, 551 (non-div) Companies)
1st Anti-Aircraft Divisional RASC (TA)
2nd Anti-Aircraft Divisional RASC (TA)
3rd Anti-Aircraft Divisional RASC (TA)
4th Anti-Aircraft Divisional RASC (TA)
5th Anti-Aircraft Divisional RASC (TA)

Appendix H

ALLIANCES WITH OTHER CORPS

In December 1920 HM King George V approved that the Army Service Corps of the Dominion of Canada, the Commonwealth of Australia, New Zealand and the Union of South Africa should be allied to the Royal Army Service Corps in recognition of their close collaboration during the Great War. The close ties forged during the years of war have been maintained ever since. The Royal Canadian Army Service Corps, throughout the 1920s and 1930s kept the Royal Army Service Corps informed of its activities, through regular contributions to the *Journal* of the Royal Army Service Corps, and kept itself fully informed of developments in the Royal Army Service Corps through contacts made between officers of the two Corps.

Through the initiative of Colonel H. C. Greer, who was Director of Supplies and Transport of the Canadian Army from 1925 to 1930, a system was arranged, by which individual RCASC and RASC officers were exchanged for a period of two years. The first RASC officer to go to Canada was Captain N. G. Duckett, who was exchanged for Major Tidswell RCASC. Captains G. E. C. Rossall, R. H. Elliot and C. H. Collinssplatt followed Captain Duckett. Lieutenant J. L. Sparling, Major P. Hennessy, Lieutenant G. E. R. Smith, Captain N. B. MacDonald and Captain D. Douglas were other RCASC officers who were attached to the Royal Army Service Corps. Four of these, Brigadier N. B. MacDonald CBE, Brigadier G. E. R. Smith CBE, Colonel D. Douglas and Colonel J. L. Sparling OBE, became Directors of Supplies and Transport in the Canadian Army and Brigadier Smith became Honorary Colonel Commandant of the Royal Canadian Army Service Corps in 1957.

A number of officers from the Dominion Forces transferred to the British Army between the wars, mainly because, owing to the small size of their own permanent forces, they saw little hope of advancement or promotion. Lieutenants A. J. Dewar (later Lieutenant Colonel A. J. Dewar DSO), G. J. Salter, H. H. Bruton (later Brigadier H. H. Bruton OBE) and G. Macleod (later Colonel G. Macleod) all came from Australia and made their careers in the Royal Army Service Corps.

Army Order 46 of 1930 announced HM the King's approval of an alliance between the Ceylon Supply and Transport Corps and the Royal Army Service Corps. To mark this alliance the Ceylon Supply and Transport Corps presented the Royal Army Service Corps with a silver table centre consisting of a double bullock cart.

In the *London Gazette* of 3 December 1932 HRH The Duke of Connaught was appointed Honorary Colonel of the Ceylon Supply and Transport Corps.

In the following year the Ceylon Supply and Transport Corps received a new title - The Ceylon Army Service Corps.

By the time that the title 'Royal' was conferred on the Indian Army Service Corps by HM The King on 3 June 1935 the Allied Corps included those of Canada, Australia, New Zealand, South Africa, India and Ceylon.

Appendix J

The Supply Reserve Depôt, Deptford

The Supply Reserve Depôt played a central part in the history of the Royal Army Service Corps supply service, and the depôt itself had a romantic past, which is not as widely known as it should be.

Before 1878 shipments of supplies for military forces overseas were made by the Admiralty (food and forage) and by the Ordnance Department (mainly fuel and light), and Deptford was the main shipping port used for this purpose.

In 1878 it was felt that there was a need for an army reserve of supplies and that the policy of depending entirely on contractors for the initial requirements of an expeditionary force should be abandoned.

A reserve of some of the essential commodities which would be required on mobilisation was established in a part of Woolwich Dockyard under the control of the Senior Commissariat Officer, Woolwich. A small staff was provided for the new depôt which was named 'The Commissariat Reserve Stores'. Labour was provided by fatigue parties from Woolwich Garrison. The formation of this unit was vindicated by the experience of the Egyptian campaigns of 1880 and 1882.

By 1888, ten years later, the Supply Reserve Depôt was a fully developed establishment, still located in a corner of the Royal Dockyard, Woolwich. It had become not only the home base for the supply of expeditionary forces but also for the maintenance of overseas stations for a large number of items. Reserves were turned over in peace time by issues to Home as well as Overseas stations.

The Depôt's first real test came with the outbreak of the South African War in 1899. It was decided that a reserve of three months supplies for the entire force should be shipped to South Africa. This amounted to a substantial tonnage: 4,000 tons each of preserved meat, biscuit and flour, 1,000 tons of jam, 750 tons of sugar, 400 tons of compressed vegetables, 110 tons of salt, 80 tons of coffee, 40 tons of tea, 7 tons of pepper and 35,000 tons each of oats and hay for the animals in the force. The receipt, inspection and despatch of these supplies was made through the Supply Reserve Depôt, which involved the repacking of a considerable quantity of stores from contractors.

In 1900 the Depôt was capable of supplying the needs of an expeditionary force of 320,000 men and 80,000 animals. This meant an average daily output of 300 to 400 tons.

In about 1903 the designation of the senior administrative officer became 'Assistant Director of Supplies'; and in 1909 this appointment was filled by Colonel S. S. Long. Three years later the administrative work of the Depôt was transferred to the War Office, and a few months later Colonel Long went to the War Office as Assistant Director of Supplies.

At the time of mobilisation in 1914 the staff of the Depôt consisted only of three officers and 35 permanent subordinates, augmented by casual labour.

When Colonel Long was still Assistant Director of Supplies at the Supply Reserve Depôt, he learnt that the Deptford Foreign Cattle Market, some four miles from the Royal Dockyard, Woolwich, had become vacant. Colonel Long immediately appreciated the possibilities of the site.

(a) there was a river frontage of 1,025 ft with a wharf and jetty. At the jetty there was 19 ft of water at low tide and 39 ft at high tide. In the inner pool seven barges could be loaded concurrently;
(b) as this was not sea-board, there were no harbour dues;
(c) the site, including adjoining areas, covered 38 acres. There were commodious storesheds;
(d) it was close to the centre of London, the principal manufacturing area for supplies;
(e) it was ideal for obtaining civilian labour;
(f) it was within easy reach of the War Office, from where the Depôt was administered by QMG 6.
(g) it adjoined the Royal Victoria Yard, which was the Royal Navy's principal victualling depôt.

The City of London Corporation had closed the Cattle Market in 1913, because frozen meat had killed the trade in foreign cattle, and were using the site as a storehouse.

At the instigation of Colonel Long it was leased from the City of London Corporation, together with the adjoining Sayes Court, part of the old Evelyn Estate, which was rented from the Estate Trustees. The whole site was purchased for the Crown by the War Department in 1925.

The Foreign Cattle Market was the site of the famous Royal Deptford Dockyard, established by Henry VIII in 1513 as the birthplace of the British Navy. Four years later a contract was placed with John Hopton for the building of a dock with gates and the necessary sluices for the construction of warships. By 1521 Deptford had become a naval station, then harbouring the *Mary Rose* (named after Mary Tudor, sister of Henry VIII) of 'the portage of 600 tons' and other naval vessels. The first Royal visit was by Edward VI in June 1549.

From here Captain Drake sailed in the *Pelican* (afterwards named the *Golden Hind*) on 13 December 1577 on his voyage around the world. On his return Queen Elizabeth I knighted him on 4 April 1581 after dining on board.

In 1588 the *Revenge* was fitted out at Deptford. She carried Sir Francis Drake's flag in action against the Armada.

The last ship launched at the Dockyard was the *Druid* — 1,322 tons — on 13 March 1869 in the presence of HRH Princess Louise and HRH The Duke of Connaught. The Royal Dockyard at Deptford proved unsuitable for the construction of ironclads, which were larger and of greater draught than wooden ships. Consequently in 1869 the yard was closed and its ship-building work transferred to Chatham. The dockyard was bought by the Corporation of the City of London and opened in its new role as a foreign cattle market on 28 December 1871.

The site leased by the War Department in 1914 included Sayes Court and an adjoining part of the old Evelyn Estate.

There are many references to the Royal Dockyard in Pepys's *Diary* and of Sayes Court in John Evelyn's diaries. Sir Walter Scott writes in *Kenilworth* of Queen Elizabeth's visit to Sayes Court.

In 1696 Admiral Benbow was tenant at Sayes Court and two years later the Czar of Russia (later 'Peter the Great') came to Deptford in order to study naval construction. Evelyn wrote: 'The Czar of Muscovy being come to England and having a mind to see the building of ships, hir'd my house at Sayes Court and made it his Court and Palace, new furnished for him by the King'. The Czar stayed for three months. Sir Christopher Wren, the King's Surveyor, estimated the damage done to Evelyn's property by the Czar and his Suite at £150. Evelyn received compensation from the Treasury. Peter the Great's visit was commemorated on a stone tablet in one of the Supply Reserve Depôt stores. The only remaining part of the mansion became the Supply Reserve Depôt Officers' Mess.

The Supply Reserve Depôt did not move to Deptford until seven weeks after the outbreak of war in 1914. As the needs of the army expanded, the 38 acres of the Cattle Market and Sayes Court sites became inadequate and it was necessary to lease other subsidiary depôts in the neighbourhood. These were Willow Walk, Deptford Wharf (with the adjoining Surrey Dock Plot), Millwall Depôt, Tower Bridge Wharf, Archangel Wharf and finally, in September 1918, Poole.

Between 1914 and 1918 the strength of the Supply Reserve Depôt increased to 95 officers, two government analysts, 294 MT personnel (military and civilian), 141 clerks, 68 superintendents and foremen, 100 gangers and checkers, 32 messengers and 2,470 labourers and packers - a total of 3,202. During the war the Depot, excluding the subsidiary depots, despatched 1,406,721 tons by sea and 931,312 tons by rail. These figures do not include despatches by barge in 1914 and 1915 or consignments by road vehicles.

The armistice in November 1918, followed by demobilisation, had an immediate effect on its activities.

The subsidiary depôts were closed in 1919. Poole was transferred to the Disposals Board in February 1920. The Depôt's own transport units were dispersed, and transport was thereafter obtained by requisition on the Officer in charge of Transport, Woolwich.

By the mid-twenties the Supply Reserve Depôt reached its final peace time establishment of five officers, one Warrant officer Class I, one government analyst and 154 civilians. These were sufficient for the handling of an average monthly tonnage of between 5,000 and 6,000 tons.

A cadre organisation was retained to allow for rapid expansion in the event of war or other national emergency. The main requirement would be an increase in manpower to deal with rapidly increasing commitments.

The system of obtaining stocks was comparatively straightforward. For the most part supplies were obtained under War Office bulk contracts, which stipulated direct delivery to the Supply Reserve Depôt. Monthly demands were submitted to the Director of Army Contracts, normally for sufficient quantities to meet issues and maintain reserves. The demands were formulated by the Supply Reserve Depôt and approved by the QMG 6 branch of the War Office.

In the Army Estimates for 1925/26 the functions of the Supply Reserve Depot were simply defined as 'the storage and turnover of war reserves of supplies, as well as the receipt, custody and issue of certain supplies for peace consumption.' That definition gave little indication of the enormous task which the Supply Reserve Depôt was called on to undertake in supplying the needs of an Army at war.

Appendix K

The RASC Record Office

On 1 July 1905 all Army Service Corps Records were collected into the Record Office at Woolwich Dockyard. The first Officer in charge of Records was Colonel F. Landon (later Major General Sir Frederick Landon KCMG CB). On 1 September 1922 the Record Office moved to Woolwich Common with the Regimental Paymaster. On 1 July 1925 the Office moved to Deptford. The move to Ore Place, Hastings took place after 1939.

Officers in charge of Records 1919 to 1939:

1918 - 1919	Colonel E. I. Ward	1926 - 1930	Colonel H. G. Reid CMG DSO
1919 - 1920	Colonel P. C. J. Scott CB	1930 - 1934	Colonel C. W. Horton DSO
1920 - 1924	Colonel R. B. Airey CMG DSO	1934 - 1938	Colonel A. M. Wilson DSO
1924 - 1926	Colonel G. H. M. Bingham CMG DSO	1938 - 1940	Colonel H. P. Raymond OBE

Appendix L

The War Department Fleet

The War Department Fleet first came into existence in 1818 when it was placed in commission under the Board of Ordnance. In 1891 the vessels were taken over by the predecessor of the Assistant Director of Military Transport, whose appointment then was Deputy Assistant Adjutant General (DAAG) for Transport. In 1908/09 certain additional vessels were taken over from the Royal Engineers.

The following are the main functions of the War Department Fleet:

(a) the provision of certain services for the Army, which cannot be readily obtained from commercial shipping sources, for example target-towing and the transport of heavy artillery equipments; (b) the transport of explosives, which otherwise could only be conveyed by land and sea at very high commercial rates; (c) the provision of water transport at remote military stations where insufficient commercial transport is available; (d) the provision of vessels to meet mobilisation requirements; (e) the provision of water transport in emergencies, and (f) the provision of vessels for the training of troops.

The establishment of the War Department Fleet in 1934 was:
steamers and steam launches 26; motor vessels and motor launches 23; barges etc 17; small boats (dinghies etc) 132.

The Fleet was distributed in a number of stations at home and abroad. The two main home stations were Woolwich and Portsmouth. Other stations were Shoeburyness, Sheerness, Dover, Devonport, Weymouth, Grimsby, Leith and the South Irish Coast Defences (Spike Island and Bere Island). Abroad there were vessels at the following stations: Gibraltar, Malta, Egypt, Singapore, Hong Kong, Bermuda and Jamaica.

All the vessels were manned by War Department civilian employees.

The Fleet was administered by the Assistant Director of Military Transport ('The Admiral of the WD Fleet'), whose office was at Woolwich Arsenal. He was directly responsible to the Director of Movements and Quartering, the War Office, (QMG 13) for the administration of the War Department Fleet, for the movement of all stores, explosives etc by War Department vessels and coastwise freighters and also for the movement of all stores etc by rail to and from all military establishments in Woolwich Arsenal, Woolwich Dockyard, Army Medical and Veterinary Stores, Woolwich Common and the Central Ordnance Depôts at Didcot, Weedon, Bramley and Purfleet. He was also available to give advice to Royal Army Service Corps Officers in charge of Transport with regard to the economical movement of large quantities of stores, explosives etc.

In 1938 a War Office Committee was appointed to examine and report on the administration of the War Department Fleet and their recommendations led to a drastic re-organisation, the Fleet then becoming the responsibility of the Director of Supplies and Transport. The re-organisation was complete by the outbreak of war in 1939.

The following Royal Army Service Corps Officers held the appointment of Assistant Director of Military Transport from 1912 to 1939:

1912 - 1914	Colonel Sir Wallis King CB CMG CVO.	1922 - 1926	Colonel G. Vawdrey CB CMG CBE.
1914 - 1918	Colonel R. A. Nugent CB.	1926 - 1930	Colonel H. M. Caddell CMG DSO.
1918 - 1919	Colonel P. O. Hazelton CB CMG.	1930 - 1934	Colonel W. T. R. Browne DSO.
1919 - 1922	Colonel E. F. Taylor CB.	1934 - 1938	Colonel C. W. Edwards DSO.
1922	Colonel L. R. Beadon CMG DSO.	1938 - 1939	Colonel T. J. R. Langmaid CVO MC.

Appendix M

The Badge and Motto of The RASC

The badge of the Royal Army Service Corps was described in Dress Regulations 1934 as follows: 'A seven-pointed star surmounted by a crown; on the star a laurel wreath; on the wreath a scroll inscribed "Royal Army Service Corps"; within the wreath a Garter with the motto "Honi soit qui mal y pense"; within the Garter the Royal Cypher.'

This badge, which dates from 1918, when the Royal title was conferred on the Corps, is an adaptation of the Army Service Corps badge which preceded it, and which itself evolved from the types of badges worn by the first Army Service Corps and the Commissariat and Transport Corps between 17 December 1869 and 11 December 1888, when the Army Service Corps, predecessor of the Royal Army Service Corps, was formed.

The badge of the newly-formed Army Service Corps was officially sanctioned by the Horse Guards authorities on 26 September 1894: 'The Badge . . . designed by Captain A. H. Terry, ASC is hereby approved as the Regimental Badge of the Corps. [signed] H. Bunbury DA for AQMG.' This badge is illustrated on p 905 of *The Records and Badges of every Regiment and Corps in the British Army* by H. M. Chichester and G. Burges-Short and differs from the 1918 Royal Army Service Corps badge in the following particulars. The Crown was St Edward's (or the Queen's) Crown. There was no scroll. The laurel wreath consisted of 33 leaves on either side. Within the garter was the ASC Monogram in place of the Royal cypher.

The badge remained unchanged until 1901 when, with the accession of King Edward VII, the crown was changed to the Tudor crown. The points of the star were less accentuated.

Between 1902 and 1918 a number of small modifications were made. The crown was lowered so that its base rested on the garter. Later, during the war, the wreath became coarser with seventeen leaves each side, instead of thirty-three, the garter tongue and buckle were modified and the letters of the monogram were broader.

The new badge of the Royal Army Service Corps, which first appeared on the cover of the *Journal* in the issue of May-June 1919, showed the new title of the Corps on a scroll, which covered a great part of the wreath, leaving only five leaves on each side. Inside the garter the monogram of the ASC was replaced by the Royal cypher G v R. A smaller Tudor crown continued to rest on the garter.

This badge continued in use until the death of King George V. In August 1936 the design of the new badge, bearing the Royal cypher of King Edward VIII, was illustrated in the RASC *Journal*. It was pointed out that in the correct design the crown should be elevated so as to be distinct from the remainder of the badge, and that the scroll should cover up to five intermediate smaller points of the star. Coloured reproductions of the badge should show the centre containing the Royal cypher as red, and the scroll containing the garter motto as blue.

The scroll containing the Corps title should not be coloured and all lettering should be in gold. There was scarcely time for the manufacturers to take note of these instructions before King Edward VIII's abdication took place and a new badge was required.

In April 1937 the design of the new badge with King George VI's Royal cypher appeared on the cover of the Royal Army Service Corps *Journal*. Apart from the cypher no changes were made.

After the 1939/45 war some further modifications were made. The crown was raised until it was above the tips of the wreath, which had smaller leaves — seven instead of five on each side — and small changes were made to the tongue and buckle of the garter. The same design was continued after the accession of Queen Elizabeth II, with the change of crown to the Queen's crown and the Royal cypher EIIR replacing that of King George VI.

An account of *The Origin and Variations of the RASC Badge* by Charles Thomas appeared in the *Bulletin* of The Military Historical Society, August 1963 (pp 4-8).

The Royal Army Service Corps motto *Nil sine labore* was adopted as the motto of the Army Service Corps in 1892 and was first used on the title page of the ASC *Journal* for August 1892. Before then 'For Queen and Country' had been used.

In the *Journal* for May 1938 the following first-hand account from Major R. H. Frost MBE was recorded:

'It is well known that Lieutenant Colonel E Grattan, the first OCASC Aldershot, was the founder of the Journal - the first issue was made in April 1891 - but before launching it he called a general meeting of representatives of all ranks, who were greatly in favour of the proposition. At a subsequent similar meeting Colonel Grattan suggested that a motto was desirable for the new Corps and recommended *Nil sine labore*. It was put to the meeting and approved.'

Appendix N

THE CORPS COLOURS

Until 1894 the Corps colours were blue and white. Soon after the formation of the Army Service Corps there was a widespread feeling throughout the Corps that these were unsuitable. This was voiced in the *Journal* of February 1894. After a discussion at an Officers' Mess Meeting at Aldershot, when it was 'unanimously passed that the colours were disapproved of', a Committee was formed with Captain P. E. F. Hobbs (later Major General P. E. F. Hobbs CB CMG) as President, to go into the whole matter, after consultation with representatives of other stations.

By September 1894 the Committee had done its work and a new pattern combining 'blue, white and golden yellow' was approved and received official sanction. The reason given was as follows: 'This departure from the former blue and white only has been made in reference to the expressed wish of a large proportion of the Corps, and the third colour is intended to represent the gold lace on the tunic and to impart character, distinctiveness and greater beauty to the colours'.

This design remained unchanged until 1911, when Lieutenant Colonel R Ford (later Major General Sir Reginald Ford KCMG CB DSO), who was secretary of the Officers' Club Committee, proposed that the tie, sash and cap should be 'of silk, in equal stripes of silver, royal blue and gold'. This was approved and the new pattern of colours accepted.

After the 1914/18 war there was once more a divergence of opinion about the colours and during 1924 correspondence was published in the *Journal*. As a result a referendum was circulated from the War Office to all Home Commands. About 90% of the officers wished the colours to remain the same, but opinions varied about what design should be adopted.

A Colours Committee was appointed and a new pattern was evolved in 1925. However it was soon discovered that the design selected was precisely the same as that already been adopted by the Old Wykehamists Cricket Club. The question was then considered at a DST Conference in 1926, and a majority vote decided to reintroduce the original three-colour design of 1894. This pattern continued unchanged until the formation of the Royal Corps of Transport, and was illustrated on page 321 of the *Journal*, August 1931. The following describes the Regimental tie: 'On a deep blue background, double white (silver) lines separated by a narrow blue line, in all ½ inch wide; each pair of white lines 1½ inches apart. Half-way between the pairs of white lines a narrow old gold line of a width similar to the blue line separating the double white lines.

The Officers' Club colours were recorded in the *Journal* of November 1921: 'New Club Colours, in blue and red, for the members of the Club have now been approved by the Committee'. They consisted of a double red stripe on a dark blue background.

In June 1930, at the Annual General Meeting of the Club the following description of the Club Colours was proposed by Captain H. R. Kerr, seconded by Major General P. E. F. Hobbs, and approved:

'(a) The Club Colours are blue and red.

(b) The Club Blazer is of blue flannel, cut double-breasted, two button front, two button cuff, lined with blue alpaca, two bottom pockets and breast pocket.

The buttons to be flat brass ones with RASC monogram engraved thereon.

(c) The Club Tie may be described as light navy blue ground with double stripes of dark crimson, ⅜ inch wide each, divided by 1/32 inch of light navy-blue, the same striping being repeated at itervals of 1⅝ inches. The ties are cut with the stripes running diagonally from the high left to the low right.'

The Corps Flag: The traditional Corps colours inherited from the Military Train were represented by a blue and white flag, before old gold was added to the Corps colours. A blue and white flag of the Army Service Corps flew at Ladysmith in the Boer war outside the headquarters of the 6th Brigade Supply and Transport Company, commanded by Captain Philip Scott, Army Service Corps.

The addition of the third colour resulted in a tricoloured flag with equal vertical stripes of blue, gold and white, arranged according to the mnemonic 'blue to the mast and white to the blast'.

During Allenby's campaign in Palestine in 1916 to 1918 the white stripe weathered badly and became undistinguishable. The colours were therefore reversed with 'white to the mast and blue to the blast'. The arrangement was subsequently adopted at Home, first at Bulford and then at Aldershot in 1921. This was confirmed by the Colours Committee set up in

1925, which ruled that 'white should be the colour nearest to the flagpole'. There was, however, still some confusion as late as December 1929 when the arrangement white/blue/yellow was published in the *Journal*. But this was quickly corrected a month later to white/yellow/blue - 'White to the mast, blue to the last', and this continued to be the official pattern until after the second world war. In 1945 another Colours Committee recommended that the normal practice of placing the lightest colour in the centre of three vertical stripes should be adopted. Accordingly it was decided to change the order of the colours to blue, white and gold and to emblazon the Corps badge on the white panel.

It was then realised that this arrangement violated the law of heraldry that two metals (in this case, silver and gold) must be separated by a colour. To correct this, the design was changed once more in 1946, and the final arrangement of the Royal Army Service Corps flag was gold to the mast, blue with the badge defacement and white, once more, 'to the blast'.

Appendix O

RASC OFFICERS' UNIFORMS

In 1919 a Royal Army Service Corps Officer's Service Dress uniform was that laid down for mounted officers, consisting of a Service Dress jacket and Sam Browne belt, khaki breeches, brown leggings and brown boots and spurs, and brown gloves. Field Officers wore riding boots instead of leggings.

In April 1923 an order was published laying down that only the following Royal Army Service Corps officers would in future wear spurs when dismounted: field officers, adjutants, officers serving with horse transport units, and officers whose duties entitled them to be mounted at public expense.

This order was simplified by another order published in June 1925, that the dress of Royal Army Service Corps officers would be: 'Service Dress uniform as laid down for mounted officers, except that officers under field rank employed on MT and Supply duties will not wear spurs'.

In November 1924 an Army Order was published, stating that a 'new Mess Jacket for Officers of the Royal Army Service Corps will be of blue cloth, with white cloth roll collar and white cloth pointed cuffs; shoulder straps of blue cloth and round points fastened with a small button; badges of rank in metal and gilt collar badges, but no buttons on the front of the jacket, and no gold braid or piping. Officers in possession of the existing pattern Mess Jacket will be permitted to continue it in wear until requiring replacement'.

This was, in fact, a reversion to the pattern of Mess Jacket worn before 1911. Between then and 1914 a new mess jacket with white silk facings on the lapels rather similar to the pattern worn by officers of the Royal Artillery was introduced. One or two officers were still wearing out this pattern of mess dress as late as 1935. The new pattern was obligatory on all officers commissioned into the Corps from 1925 onwards.

In January 1936 a new pattern cane was introduced. In place of the thin whippy 'swagger' cane previously carried, a 'Rattan cosh' was to be carried. This was described as a 'polo stick end, 2 feet in length and ⅔ inch in diameter, not covered with leather'.

In December 1936 a new pattern field boot (the Royal Artillery pattern boot) was introduced to replace leggings and ankle boots.

In August 1937 the coloured field service cap was first authorised.

In July 1938 authority was given for the wearing of Service Dress trousers on training. At the same time it was laid down that the frog would only be worn on the Sam Browne belt when a sword was carried.

In October 1938 breeches and field boots were replaced by Service Dress trousers.

Appendix P

THE ROYAL TOURNAMENT

The Army Service Corps was closely involved with the Royal Tournament from its earliest beginnings and throughout its whole history has provided a large proportion of the managerial and adminstrative staff.

Although 1880 was the year from which the Tournament dates its existence as a permanent institution, two years earlier the Volunteer Force gave a military display on Wimbledon Common which sowed the seeds of later developments.

The first performance in 1880 was staged at the Royal Agricultural Hall, Islington, as a private enterprise by Volunteer Officers, with the object of raising funds for the Duke of Cambridge's Fund for Old and Decayed Soldiers. Unfortunately it failed in its object and ended up with a deficit of £500. However, by including events with a more popular appeal, the organisers managed to attract greater public support and to improve their balance sheet, so much so that the Regular Army became interested, and eventually assumed responsibility for the Military Tournament, which later, with the co-operation of the Royal Navy, became the Naval and Military Tournament. The first performance of the Military Tornament took place at the Agricultural Hall in 1895.

The Tournament remained at the Agricultural Hall until 1906, when it was transferred from Islington to Olympia. Its objects continued to be to raise money for Service charities. In 1928 its profits reached a record figure of £27,000.

When the Royal Air Force was formed all three Services co-operated in the organisation and presentation of the re-named Royal Tournament.

Colonel Sir Edward Ward was Honorary Secretary of the Military Tournament from 1895 until 1899, and was mainly responsible for its early development. He had, as Honorary Treasurer, Captain C. W. King ASC who, for more than forty years, continued to devote his energies to the management and direction of one of London's most successful and spectacular shows. In 1911, as a Colonel, he became Honorary Secretary and Manager. When the Tournament was resuscitated after the Great War, Brigadier General Sir Charles Wallis King KCVO CB CMG, having retired from active service, continued to be responsible for all the administrative arrangements until an official secretary was appointed in 1933. He then became Chairman of the Tournament, in which position, freed from the ties of detailed administration, he could devote himself to the higher direction of the Show.

His son, Captain C. W. C. King, who was also a regular officer in the Royal Army Service Corps, joined the Arena and Display Staff in 1920 and remained in that capacity until 1933, when he retired from the Army and was appointed the first official Tournament Secretary.

In 1936 there were seventeen serving or retired members of the Corps on the administrative staff; four in the Treasury Office, four in the Box Office, three in the Secretary's Office, two in the Press Office, one in charge of the Arena and Displays, two in charge of Arrivals and Departures and one in the Quartermaster's Stores.

Lieutenant Colonel A. F. Gleeson OBE controlled the Box Office, with the assistance of Captains A. V. Burd and C. Collier and Lieutenant (later Lieutenant Colonel) F. H. Giles OBE, a most important office, requiring endless tact and diplomacy.

Major T. G. Spain was one of the longest serving members of the team, having started his connection with the Tournament in 1898. As a Warrant Officer he was one of the Box Office staff and later took over the Treasury office, until his death in 1938. In the Secretary's Office Captain King was assisted by two well-known retired Staff Sergeant Majors, W. J. Duffin and V. C. Soggee, both of whom joined the Staff, while they were still serving in the Corps, in 1907.

Major T. Sinfield, who distinguished himself as a Corporal in the mounted events in the 1890s, was for many years in charge of the Arena and Displays, and could always call on the Royal Army Service Corps Riding Esablishment, Woolwich, for his arena parties, until the Riding School moved to Aldershot.

Apart from the administrative staff at Olympia there were many others who contributed greatly to the success of the Tournament. Staff Sergeant Major (later Lieutenant (QM)) A. W. Flood was for a long time connected with the supply of forage and later became an Assistant Treasurer. Major J. R. Boxall MBE DCM TD began his connection with the Royal Tournament in 1910 and, during a long career assisted at forty-five performances. After retiring as a WO1 (1st Class Staff Sergeant-Major) in 1936, he was commissioned as a Captain in the Territorial Army in 1939 and served in Anti-Aircraft Command throughout the war. He continued to support the Tournament until his final retirement in 1965.

The Corps always took a prominent part in the mounted events and in the displays. The Waggon Dismantling competition and the Ride and Drive Display were popular events before the 1st World War. In 1923 an exhibition of Tandem Driving was staged by Royal Army Service Corps, Woolwich, and in the following year the Corps display was witnessed by HM the King and the team personally congratulated by HRH the Duke of Connaught, Colonel of the Royal Army Service Corps.

While Captain J. R. Allen was equitation officer at Woolwich, with Staff Sergeant (later Major) 'Tim' Pearson as Riding Master, he was indefatigable in his efforts in support of the Tournament. Not only was he one of the most accomplished performers in the mounted competitions, but he was the inspiration behind the Corps displays and the immaculate Royal Army Service Corps Stables, which were always one of the most popular side-shows in the 1920s. Later, when he was at Feltham, he continued his support of the Tournament in the preparation of the Feltham Tug-of-Wag teams.

With the disappearance of horse transport the Royal Army Service Corps was forced to switch the emphasis of displays to the less glamorous motor transport. In 1933 the Corps staged a most successful display entitled 'The Transport Cavalcade 1799-1933'. At that time the Royal Tournament still took place in the last week of May and the first in June.

After the 1939-1945 war the Horse Transport Training Company Royal Army Service Corps at Aldershot continued to make its contribution to the Royal Tournament, when it moved to Earls Court. Among other successes Major G. G. R. Boon MBE won the King's Cup in 1949. In 1962 Major T. J. Brown won the King's Cup on 'Finisterre' and in 1963 the Company staged a mounted Display which included tent pegging, riding in tandem and vaulting.

It has been the practice for the Head of the Corps to be a member of the Committee of the Royal Tournament and to do duty in the Royal Box.

Another important link was established when Major General W. H. D. Ritchie CB CBE on retirement from the Army in 1957, became Managing Director of Earls Court.

Appendix Q

CORPS MEMORIALS

St George's Garrison Church, Aldershot

Most Royal Army Service Corps officers who served at Aldershot between 1919 and 1939 will have taken part in a Church Parade and Service at St George's Garrison Church. Whatever one's personal feelings about Church Parades may be, there is no doubt that these were impressive ceremonies and the crowded Garrison Church echoing the soldiers' singing had a spiritual atmosphere, which brought inspiration to many of those who were present.

St George's Church was built as a replacement for the previous 'Iron Church' and was completed in 1893. Its foundation-stone was the only church foundation-stone laid by Queen Victoria during her whole reign. The Church has often been affectionately referred to as the 'Corps Cathedral'. The Church Service on the Sunday of Corps Week, followed by the March Past and the address by the Representative Colonel Commandant has provided a fitting climax to the week's programme.

The Church is in itself a memorial to those of the Corps who have died for their country, for it contains many of the memorials erected in memory of past members of the Royal Army Service Corps and its predecessors.

The earliest dates back more than a hundred years, and consists of a brass tablet on the west wall in memory of the members of the Commissariat and Transport Corps who died in the Egyptian campaign of 1882. The tablet contains 51 names. Next comes the 'Barnes Memorial' in the south aisle, commemorating those who fell in small wars between 1884 and 1898. There are six tablets, containing 66 names of men who fell in battles of the Nile, Suakim, Ashanti, Dongola, Mashonaland, Crete and the West Coast of Africa.

The pulpit was given in memory of Assistant Commissary General R. H. Powell, who died in 1884.

These three memorials were oiginally in the Iron Church.

Three windows in St George's are in memory of officers who served in the Army Service Corps — Colonel G. H. Jessop, Lieuteant Colonel B. Haygate and Lieuteant Colonel W. A. F. Williamson, erected by their brother officers and comrades.

The east window and the reredos were erected as the Corps memorial to those who lost their lives in the South African War. The 342 names are engraved on the panels at the back of the sedilia on the south side of the sanctuary. The unveiling was carried out by Lieutenant General Sir John French on 11 April 1903.

The Great War Memorial consists of the mural mosaics on the north and south sides of the altar. The Chancel also holds the Books of Remembrance, containing the names of 8,479 officers and men of the Corps who gave their lives between 1914 and 1919, those who died in action in the 1939 - 1945 war and those killed in action between 1919 and 1939 and 1946 onwards.

A set of marble tablets placed on the walls of the Chancel and Sanctuary commemorate the names of 106 officers who died between 1907 and 1937. An extension to this memorial was erected during 1937 and 1938 on the south-east wall of the nave, and dedicated by the Chaplain General on Corps Sunday in 1938. The extension consists of a marble and stone memorial with twenty panels on which the names are inscribed annually, according to the wishes of relatives. The memorial bears the following inscription: 'To the Glory of God and in memory of the Officers of the Royal Army Service Corps whose names are inscribed hereon'.

The memorial for those who fell in the Second World War consists of the communion rails and a tablet let into the floor of the chancel. The rails are of carved oak. The sturdy nature of the rails is relieved by a carved trail and a decorated Corps badge as the centre motif, all below the hand-rail. The decoration includes the letters RASC and the emblems of the British Isles, the rose, thistle, daffodil and shamrock form the background for the badge. The centre opening is seven feet wide, and filled by two gates of unequal size. The arms of the four Patron Saints, St George, St Andrew, St Patrick and St David are on the outer standards. The two centre shields show the Greek symbols for Jesus and Christ.

There are also: a brass plate on the wall of the south aisle in memory of Lieuteant Colonel Hugh CD Huxham, a marble plaque, which was originally in the Garrison Church at Woolwich, commemorating 21 officers, nine warrant officers and 347 NCOs and men who fell in the South African War, three brass plates on the wall of the south aisle, originally in St George's Church, Singapore, in memory of 23 NCOs and men of 55 Air Despatch Company Royal Army Service Corps, killed by enemy action in Malaya, and a stone plaque in the floor of the chancel in memory of Brigadier W. N. White CB DSO.

When the Roman Catholic Garrison Church of St Michael and St Sebastian was closed and St George's Garrison Church became the Church of St Michael and St George in January 1973, the following memorials were erected:

Brass tablet and seven stations of the Cross on the south wall, erected by the Roman Catholic officers of the Royal Army Service Corps in memory of the officers and men who died in the war of 1939-1945, and removed from the Roman Catholic Church: a brass tablet in the porch, commemorating the transfer of the Church to the Roman Catholic Military Authorities and welcoming progress towards unity between the two churches.

Royal Corps of Transport Memorial Chapel on the south side of the Chancel: the following inscription is on a brass tablet on the exterior — 'This Memorial Chapel of the Royal Corps of Transport and its predecessors, which was dedicated on Corps Sunday 1974, was created to replace that previously located in the Roman Catholic Garrison Church of St Michael and St Sebastian, Aldershot, following the closure of that Church. It contains the Memorials translated from that Chapel which were erected by offiers of the RCT/RASC to the memory of the fallen and in commemoration of the return of the Corps to their home in the new Buller Barracks in 1970.'

The War Memorial

The Royal Army Service Corps War Memorial in Buller Barracks was unveiled on 29 April 1923 by Lieutenant General Sir Frederick T. Clayton KCB KCMG. He was supported by the two Commanding Officers at Aldershot, Lieutenant Colonel R. C. Jellicoe CBE DSO, Service Companies, and Lieutenant Colonel C. R. T. Annesley DSO, No 1 Corps Depôt. The Memorial was dedicated by the Assistant Chaplain General, The Rev F. I. Anderson CMG Hon Chaplain to the King.

The Collect

The special collect of the Royal Army Service Corps includes references to phrases in the Bible which have a special significance for the Corps:

'Almighty God, Who didst carry Thy people all the day of old (1), and Who dost supply all our needs (2), uphold us who serve Thee in the Royal Army Service Corps in all our works according to Thy Will, that by well doing as good and faithful servants we may enter into Thy joy (3), through Jesus Christ our Lord.'

Notes: (1) Isaiah lxiii, 9: 'He bare them, and carried them all the days of old.'
(2) Philippians iv, 19: 'My God shall supply all your need.'
(1) and (2) Compare 'Supply and Transport.'
(3) Matthew xxv, 21: 'Well done, thou good and faithful servant . . . enter thou into the joy of the Lord.'

Appendix R

MISCELLANY

Reindeer Transport: In the early part of 1919 a Reindeer Transport Corps was formed in North Russia under the command of Captain Squarey. It consisted of 2,000 reindeer and 600 sleighs and 500 Lapps, who were mostly owners of the animals and experts at looking after them and driving the teams. A team normally consisted of three reindeer with a fourth secured to the rear of the sleigh to act as a brake on downhill stretches. Each sleigh carried a load of about 600 lb. This form of transport proved invaluable in areas around Archangel and Murmansk, where other forms of transport could not be used. On one occasion a battalion of infantry was successfully lifted over a distance of more than 100 miles.

Royal Army Service Corps, Feltham: Feltham first became a Royal Army Service Corps station in 1922 and was developed as the main Workshop and Driving School of the Corps, with a Heavy Repair Shop (HRS), a Vehicle Reserve Depôt (VRD) and Mechanical Transport Stores Depôt (MTSD) and 'R' Depôt Company (the Driving School). Feltham was the first 'horseless' Royal Army Service Corps station.

Royal Army Service Corps Headquarters Officers' Mess, Buller Barracks: On 6 October 1923 the approval of the Army Council was given for the Officers' Mess at Aldershot to be recognised as the Headquarters Mess of the Corps. The Mess was completed in 1895 and Captain P. E. F. Hobbs was the first PMC (President of the Mess Committee). The 50th Anniversary was celebrated on 30 January 1945, fifty years to the day after the celebration of the opening of Buller Mess. Fifty seven officers were present and the Corps band played selections from the programme of music played fifty years earlier.

In the collection of Mess plate are two impressive pieces of silver - copies of the famous 'Warwick Vase' - which were presented to the Regular Officers of the Corps by the temporary officers, who had served in the 1914 - 1918 war. The presentation was made at the end of 1918 at Bulford which, at that time, was one of the main Army Demobilisation Centres. The Army Football trophy is also a replica of the 'Warwick Vase'.

A fine mahogany dining table in Buller Mess was presented to the Officers of the Royal Army Service Corps by Major General Sir Richard Ewart KCMG CB CIE DSO ADC in 1926. He had a distinguished career in the Indian Army and became Colonel Commandant of the Indian Army Service Corps in 1925. The table had at one time belonged to Skinner's Horse.

It was a practice during the 1920s and early 1930s for members of Junior Officers' Courses to present a silver goblet on completion of their course. By 1939 there were sufficient of these goblets for one to be set in the place of every officer dining in the Mess.

During this period there were four formal dinner nights in each week, at which dining members wore mess dress. Wednesdays, Saturdays and Sundays were supper nights.

Resident members of the Mess could opt out of dinner and have an early supper on dinner nights, but were required to attend at least two dinner nights in each week. When the old Mess was demolished to make way for the new Mess, which was completed in 1969, a number of the original oak panels from the walls of the Mess Room were made into coffee tables by Colonel J. R. Burgess TD, at that time Regimental Secretary, who was an accomplished craftsman.

Royal Army Service Corps in Aden: Until the end of 1928 the administration of Aden was the responsibility of the Indian Army. There was a small combined detachment of the Royal Army Service Corps and the Indian Army Service Corps to provide the necessary services for the Garrison. When the administration was transferred to the Royal Air Force in 1928, Aden ceased to be a Royal Army Service Corps station.

Royal Army Service Corps Locomotive: On 14 January 1936 Major General W. K. Tarver CB CMG, Representative Colonel Commandant, officially named Locomotive No 6126 *Royal Army Service Corps*. The engine was a 'Royal Scot' Class locomotive of the London Midland and Scottish Railway and made its first official run from Euston to Crewe.

BIBLIOGRAPHY

The Journal of the Royal Army Service Corps

The Royal Army Service Corps *Quarterly*

Issues of the War Office List

The Royal Army Service Corps. History of Transport and Supply in the Army Vols I and II Sir J. W. Fortescue and Colonel R. H. Beadon. Cambridge University Press. 1930 and 1931

The Story of the Royal Army Service Corps 1939 - 1945. The Institution of the Royal Army Service Corps. G. Bell & Sons Ltd. 1955

The Story of the Royal Army Service Corps and the Royal Corps of Transport 1945 - 1982. Editor in Chief: Brigadier D. J. Sutton OBE. Leo Cooper and Secker and Warburg. 1983

Citizen Soldiers of the Royal Engineers Transportation and Movements and The Royal Army Service Corps. 1859 to 1965. Colonel G. Williams OBE. Insitution of the Royal Corps of Transport

Wait for the Waggon. The story of the Royal Canadian Army Service Corps. Arnold Warren. McClelland and Stewart Ltd. 1961

A Short Account of Canteens in the British Army. The Hon Sir John Fortescue. Cambridge University Press. 1928

Imperial Military Geography. Brigadier D. H. Cole OBE. Sifton Praed & Co Ltd. 11th Edition. 1953

The British Overseas. C. E. Carrington. Cambridge Unversity Press. 1950

The Royal Army Service Corps. Lieut Colonel Graeme Crew. The Famous Regiments Series edited by Lieut Gen Sir Brian Horrocks. Leo Cooper Ltd. 1970
Afghanistan. 1919. Lieut General G. N. Molesworth. Asia Publishing House. 1962
North-West Frontier. People and Events 1839-1947. Arthur Swinson. Hutchinson of London 1967
The Frontier 1839-1947. Major J. C. Elliott. Cassell. London 1968
Official History of Operations on the North-west Frontier of India. 1920-1935. Government of India Press, Delhi. 1945
Afghanistan. Lieut General Sir George MacMunn. Bell. 1929
The Romance of the Indian Frontiers. Lieut General Sir George MacMunn. Cape 1931
Bugles and a Tiger. John Masters. Michael Joseph Ltd. 1956
This was my Life. Lieut Colonel Geoffrey Barnett. Printed privately. 1981
Tales of the Sudan Defence Force. Editor, Colonel John Orlebar. Printed privately. 1981
Archives of the Royal Army Service Corps and Royal Corps of Transport.

List of Abbreviations

AA	*Anti-aircraft*
AA & QMG	*Assistant Adjutant and Quartermaster General*
AAG	*Assistant Adjutant General*
AASC	*Australian Army Service Corps*
ADGB	*Air Defence of Great Britain*
ADMT	*Assistant Director of Military Transport*
ADST	*Assistant Director of Supplies and Transport*
AFV	*Armoured Fighting Vehicle*
AG	*Adjutant General*
AQMG	*Assistant Quartermaster General*
ASC	*Army Service Corps*
Brig	*Brigadier*
Brig Gen	*Brigadier General*
Capt	*Captain*
CASC	*Canadian Army Service Corps or Ceylon Army Service Corps*
CIA	*Chief Inspector of Armaments*
CIMT	*Chief Inspector of Mechanical Transport*
C-in-C	*Commander-in-Chief*
CO	*Commanding Officer*
Cpl	*Corporal*
CQMS	*Company Quartermaster Sergeant*
CRASC	*Commander Royal Army Service Corps*
CSD	*Command Supply Depôt*
CSM	*Company Sergeant Major*
DAA & QMG	*Deputy Assistant Adjutant and Quartermaster General*
DAAG	*Deputy Assistant Adjutant General*
DADST	*Deputy Assistant Director of Supplies and Transport*
DAQMG	*Deputy Assistant Quartermaster General*
DDST	*Deputy Director of Supplies and Transport*
DQMG	*Deputy Quartermaster General*
DST	*Director of Supplies and Transport*
Dvr	*Driver*
EFI	*Expeditionary Forces Institute*
FM	*Field Marshal*
Gen	*General*
GHQ	*General Headquarters*
GOC	*General Officer Commanding*
HMT	*Hired Military Transport*
HP	*Half Pay*
HRS	*Heavy Repair Shop*
HT	*Hired Transport or Horse Transport*
IASC	*Indian Army Service Corps*
KAR	*Kings African Rifles*
LCpl	*Lance Corporal*
L of C	*Lines of Communication*
Lieut or Lt	*Lieutenant*
Lieut Col	*Lieutenant Colonel*
Lt Col	*Lieutenant Colonel*
Lieut Gen	*Lieutenant General*
Maj	*Major*
MAC	*Motor Ambulance Convoy*
Maj Gen	*Major General*
MASC	*Malayan Army Service Corps*
MRU	*Mobile Repair Unit*
MT	*Mechanical Transport, Military Train or Military Training*
MTSD	*Mechanical Transport Stores Depôt*
NAAFI	*Navy Army and Air Force Institute*
NCO	*Non-Commissioned Officer*
OC	*Officer Commaning*
O i/c	*Officer in charge*
OR	*Other Rank*
psc	*Passed Staff College*
POL	*Petroleum, Oil and Lubricants*
Q Branch	*Quartermaster General's branch*
QM	*Quartermaster*
QMG	*Quartermaster General*
Q (Mov)	*Q Branch concerned with Movements*
RAASC	*Royal Australian Army Service Corps*
RAMC	*Royal Army Medical Corps*
RAOC	*Royal Army Ordnance Corps*
RASC	*Royal Army Service Corps*
RCASC	*Royal Canadian Army Service Corps*
R Signals	*Royal Corps of Signals*
RCT	*Royal Corps of Transport*
RE	*Corps of Royal Engineers*
RHQ	*Regimental Headquarters*
RIASC	*Royal Indian Army Service Corps*
RNZASC	*Royal New Zealand Army Service Corps*
RSM	*Regimental Sergeant Major*
RUSI	*Royal United Services Institution*
RWAAF	*Royal West African Frontier Force*
SAA	*Small Arms Ammunition*
Sgt	*Sergeant*
S and T	*Supplies and Transport*
SQMS	*Staff Quartermaster Sergeant*
SRD	*Supply Reserve Depôt*
S Sgt	*Staff Sergeant*
SSM	*Staff Sergeant Major*
VRD	*Vehicle Reserve Depôt*
WD	*War Department*
WO	*Warrant Officer*
WO 1	*Warrant Officer Class 1*
WO 2	*Warrant Officer Class 2*

Index of Names

Figures in *italics* refer to illustrations

INDEX OF UNITS

Figures in *italics* refer to illustrations

GENERAL INDEX

Figures in *italics* refer to illustrations

SUBSCRIBERS

Presentation Copies

I Her Majesty the Queen
II HRH Princess Alice, the Duchess of Gloucester GCB CI GCVO GBE
III Royal Australian Corps of Transport
IV Royal New Zealand Corps of Transport
V Army Service Corps of India
VI Army Service Corps of Pakistan
VII The Sri Lanka Army Service Corps
VIII The Malaysian Service Corps
IX The Worshipful Company of Carmen
X The Institution of the Royal Corps of Transport
XI Field Marshal the Lord Carver GCB CBE DSO MC

1 Maj Gen P.G. Turpin CB OBE MA FCIT
2 Clive Birch FSA FRSA
3 Maj Gen P.H. Benson CBE FCIT
4 Maj Gen P. Blunt CB MBE GM
5 Maj Gen F.J. Plastkett CB MBE
6 Maj Gen D.H. Braggins CB
7 Maj Gen D.B.H. Colley CB CBE
8 Maj Gen C.E.G. Carrington CBE
9 Sidney Atkins
10 Lt Col M.H.G. Young
11 Col E.H. Reeder
12 Lt Col B.C. Vaughan
13 Capt K.M. Davidson TD
14 Brig R.A. Nightingale MBE
15 Maj Gen P.F. Claxton CB OBE MA
16 Brig B.G.E. Courtis
17 Lt Col G.T. Pearce MBE
18 R.R. Smith
19 Capt J. Venmore
20 Robert Hughes
21 Lt Col L. Huxtable
22 Lt Col Lincoln Jones MBE
23 Lt Col J.R. Evans
24 Maj T.A. Riddle ERD
25 Brig W.R. Barker CBE
26 Lt Col P.M. Edgell OBE
27 R.W. Munro
28 Brig A.K. Dixon
29 J.C. Atlee
30 Maj (QM) J. Perrins
31 Brig W.M.E. White CBE
32 Brig P.I. Palmer OBE ADC
33 Brig M.W. Betts
34 Maj Gen V.H.J. Carpenter CB MBE
35 Lt Col D.R. Chinn
36 Lt Col I.H.W. Bennett
37 Brig D.J. Sutton OBE
38 Capt G. Russell
39 44 The School of Transportation & Prizes, Aldershot
45 Maj W.K. Jewell
46 Lt Col J.J. Mallon
47 Maj A.R.J. Batty MBE
48 Prof R. Graveson CBE QC
49 Lt Col O.T. Hall
50 Maj Gen Sir John Potter KBE CB
51 Capt T. Harrison MBE OStJ TD
52 Capt J.A.S. Downes BA
53 Allhallows School
54 Bedford Modern School
55 Cranbrook School
56 Dean Close School
57 Denstone College
58 Durham School
59 King's College, Taunton
60 Lord Wandsworth College
61 Pocklington School
62 Taunton School
63 West Buckland School
64 Lt Col C.J. Duxbury MBE
65 Hon Librarian, Institution of the RCT
66 Lt Col T.A. Danton-Rees
67 Maj M.D. Isherwood
68 Maj Gen J.E. Witt CB CBE MC
69 Col H.B. Cox
70 Maj D.G. Balcombe
71 Brig R.E.L. Jenkins CBE
72 Brig E.W.T. Darlow OBE MA
73 Brig R.M. Bullock
74 WOs & Sgts Mess, Duke of Gloucester Barracks
75 Officers' Mess, 29 Tpt & MOV Regt RCT
76 Col J.A. Butler MC TD
77 Officers' Mess, 157 (Wales & Midland) Tpt Regt RCT(V)
78 Capt R.A. Dixon-Warren
79 Maj Gen W. Bate CB OBE DL FCIT
80 Brig C.J. Carey
81 Brig R.G. Harmer
82 84 WOI (RSM) R.L. Wilson
85 SSgt R.E. Wallis
86 Maj W.P.B. Moss
87 Maj J. Walker
88 89 Maj M. Swale
90 J.F.L. Lowndes
91 Lt C.P. Osbourn
92 Col R.C. Atwill
93 Maj T.M. Thompson
94 Lt Col E.S. Yates
95 PRI 60 Sqn RCT
96 Maj P.G. Malins MBE MC
97 Brig E.V. Molyneux OBE MC
98 A.G.F. Pallister
99 Jonathan Kimpton
100 Maj R.J. Carter
101 Brig J.D. Lofts MBE
102 Ian W. Fawkner
103 Maj K.B. Wilson
104 Maj J.G. Anderson
105 Col E. Denis Higgs ERD
106 Lt Col L. Cromwell OBE
107 Maj R.C. Blunt MBE
108 Capt M.B. Grey
109 Capt J.H. Wyss MBE
110 Lt Col G.C.E. Crew OBE BA
111 Lt Col Grattan Hart DL
112 Lt Col D. Boot
113 Lt Col D.R. Black
114 WO1 (RSM) Harvey
115 Maj G.O. Davies MBE
116 Maj J. Howard
117 Lt Col C.A. Murray
118 WO2 Turner
119 Maj D.J. Norton
120 Brig W. Horsfall OBE
121 Maj Peter Burr
122 Maj (QM) D.I. Weir
123 G.N. Delaney
124 Maj R.H. Dobson
125 Maj T.C. Thornton
126 Maj T.W.F. Odell MBE
127 Brig R.G. Thorne OBE
128 Maj C. Marder
129 Lt Col J.F. Hallum
130 Capt D.J. Shouesmith
131 Officers Mess, 1 Armoured Division Transport Regt RCT
132 Officers Mess, 153 (Highland) Arty SP Regt RCT(V)
133 WOs & Sgts Mess, 153 (Highland) Arty SP Regt RCT(V)
134 Col C.A. Butterfield
135 Lt William James Parrott
136 Lt Col W.J. Heaps
137 38 Sqn RCT
138 Brig J. Heptinstall
139 Maj J.E. Willis
140 Maj Gen D.H.V. Buckle CB CBE
141 Col J.F.E. Pye OStJ
142 Lt Col A.R.M. Adams MBE
143 Col R.J. Boyles
144 Col R.G. McCoy ERD
145 Capt B.L. Tarr
146 Maj G.C. Refoy
147 Maj F.T. Richards
148 Lt Col S. Hubbard OBE
149 Lt Col R.H.F. Burrell
150 Maj J.A. Northover
151 Maj (QM) D. Morley MBE
152 Lt Col M.C. Sims
153 Maj G.H. Atkinson

154 Maj A.H. Larsen
155 Maj R.A.D. Sinclair
156 Capt M.J. Gartside
157 Brig J.B. Sandberg CBE
158 Capt I.M. Lawrie
159 Brig A.J. Simmons
160 Maj (QM) H.M. Whitehead
161 Capt (QM) M. Bradley
162 Lt Col E.A. Kynaston
163 Maj W.A. Teague
164 Col J.M. Kelsall
165 P.R.L. Lee-Emery
166 Maj B.T.C. Smith
167 Lt Col H.T. Abbott
168 Brig E.R. Goode CBE
169 Lt Col R.C. Smith
170 Col R.L. Wallis
171 Capt C.A. Platt
172 Lt Col J.H. Walters OBE
173 Lt Cdr J.J. Town-Clear
174 Lt Col S.T.M. Pledger
175 Capt N.C. Sykes TD
176 Lt Col B.B. Bateson
177 Lt Col J.G. Hambleton MBE
178 Col E.E. Bonner TD
179 Maj H.E.S. Hayward
180 Maj Terry Lees
181
182 Lt A.I. Griffiths
183 Lt Col M.F.I. Cubitt MBE
184 Lt Col P.M. Carolan
185 Lt Col D.B. Green
186 54 Engineer Support & Ambulance Sqn RCT
187 Lt Col A.T. Hockey ERD
188 Maj M.G. Lambert
189 Col B. Mitchell
190 Maj R.H.G. Barton
191 Brig J.D. MacDonald CBE
192 Maj M.A.A. Smith
193 Lt Col R.W. Armstrong
194
195 Capt A.J. Paterson
196 Lt Col A.R. Skipper
197 151 (Greater London) Tpt Regt RCT(V)
198 Lt Col D.L. Merrylees
199 Lt Col R.J. Royle BA
200 Col Sir David Cooke Bt BA
201 Maj A.J. Maddern
202 Maj H.B. Stevens
203 Maj P.S. Reehal
204 Maj M.S. Rees
205 Capt P.T. McGrath
206 Lt Col H.L. Dalwood
207 Brian S. Bartlett
208 Maj I.J. Tucker
209 Col P.F. Hollins
210 Lt Col H.I. Charkham OBE
211 Maj W. Dow
212 Brig T.A.K. Savage MBE
213 Capt J.A.M. Smith
214 Col F.H. Potter OBE
215 B.B. Rathbone
216 Lt Col D.A. Mead TD
217 Maj H.T. Reoch
218 Lt Col J.P. Shoesmith TD
219 Lt Col F.O. Atkinson
220 Capt D.A. Wells
221 Maj V. Sugden
222 Brig G.E.C. Rossall CBE
223 Lt Col J.O. Penrose MBE TD
224 Lt Col B.L. Barrett
225 Lt Col R.W. House
226 Lt Col K.C. Brown TD
227 Col S.H. Spackman TD
228 Maj H.J. March
229 Maj D.G. Chick
230 Col J. White OBE
231 Capt P.E. Alberry
232 Maj Douglas
233 Col R.A. Gill TD JP DL
234 Capt N.G. Campbell
235 Capt J.H.W. Beardwell
236 Lt Col P.L. Morgan
237 Lt Col S.H. Scroope
238 Maj C.W.E. Collins
239 Capt W. Tate
240 Maj H.J. Ford
241 Maj J.M. Lilley
242 Lt Col C. Adwick OBE
243 Lt Col A.J.M. Smetham
244 Col H.S. Butterworth MBE MC TD
245 Maj T.H. Garner
246 Lt Col B.C. Garrard
247 Lt Col A.R. Allum
248 Brig K.C. Davis MBE
249 Col C.G. Dickie TD DL
250 Col R.F. Discombe OBE
251 Maj R.J. Selley
252 Lt Col B.W. Ballinger
253 P.J. Ballinger
254 Col B.G. Jones ERD TD
255 Maj Gen I.S. Baxter CBE
256 Capt A.J. Bray
257 Lt Col P.K. Robinson
258 Maj R.M. Taylor
259 Maj M.H. Soulsby
260 Lt Col C.D. Alexander
261 Maj N.G. Vyvyan
262 K.P. Haselock
263 Capt P.I. Webb TD
264 Lt D.R. Woodford
265 Lt Col P.W.M. Roberts
266 Capt P.J. Brown
267 Maj E.B.D. Mitchell TD
268 Capt L.A. Palmer TD
269 Maj D.J. Pleasants
270 A.G. Jackson
271 John Boxall
272 Capt M.J. Boden LLB
273 Lt Col M.R.U. McCartney
274 Capt S.P. Cowlam
275
276 Col G.J. Yeoman MBE
277 Brig W. Grierson MBE MM
278 Leo Cooper
279 M.SSgt William H.J. Jenkins
280 Maj D.J. Owen
281 W.E.J. Henderson
282 Lt Col R.R.L. Thomas PhD DBA
283 Clifford Wilson
284
285 Museum of Army Transport
286 A.C. Pearce
287 Sgt A.P. Lambert
288 Col D.W. Ronald
289 Lt Col R. Millington
290 Lt Col G.J. Shawley TD
291 Capt K.J. Wall
292 Maj R.M. Rodgers
293 Lt Col G.C.E. Crew OBE
294 Col W.H. Slack ERD DL
295 W.K. Adrian
296 Col J.S. Riggall MBE
297 Maj A.F. Neve
298 Maj J.H. Innes TD CA
299 Annabel M. Turpin
300 John Wilmshurst
301 Maj Paul R. Sheridan TD BA
302 Lt Col L.H. Smith OBE TD DL
303
304 J.M.B. Gotch
305 Reginald S. Rowland
306 R.W. Travis
307 Mrs S.G. Haswell
308 Meg Taylor
309 W.A. Morris
310 Norman Henry Collins
311 Lt Col J.F. Waterfield
312 Maj R.M. Aickin
313 Lt Col L. Clifford
314 Douglas M. Gluckstein
315 Col T. Lord CBE ERD TD
316 Maj G.W. Murley
317 Maj Gen J.R. Reynolds CB OBE ERD DL
318 Capt P.J. McGuigan
319 Frederick S. Eckett
320 Maj C.W.P. Coan
321 Lt Col (SQM) T.E. Nelson
322 T.L. Jose JP
323 O. Sunderland TD
324 Michael D. Kinsley
325 Maj E. Hughes
326 Lt Col P. de L. Bainbridge
327 Lt Col M.G. Wellings
328 Glyn Samuel
329
330 Major M.J. Dunn
331 Col W.E.J. Allen
332 Maj Gen E.H.G. Lonsdale CB MBE MA
333 Maj (QM) W.A.C. Tout
334 Lt Col C.G.M. Marsh
335 Maj R.C. Croslegh
336
337 RCT Museum Archives
338 Brig D.N. Locke OBE
339 Col A.R. Tapp
340 Maj A. Harding
341 Col J.C. Lucas OBE
342 Col J.S.M. Walker
343 Lt Col R.K. Cooley
344 Lt Col D.S. Wooles
345 Col J.D. Payne
346 Rt Hon Viscount Downe
347 Brig N.I.B. Speller MBE
348 Lt Col D. Neighbour
349 Lt Col J.W. Clement OBE TD
350 Rt Hon Sir Edward Du Cann KBE
351 Capt R.J. Smith
352 Lt R.A. Doughty
353 Lt Col P.C. Durbin OBE TD
354 Colin M. Edwards
355 Lt Col A.L. Bridger OBE
356 Maj C.H. Maginniss
357 Col A.G. Bell
358 Maj M. McGuire
359 Lt Col J.J. Treasure
360 Coventry Branch RASC/RCT Association
361 37 Sqn RCT
362 Lt T.J. Higgins
363 Maj P. Warwick
364 Cllr W. McKernan
365 14 Sqn RCT
366 Col P.E.B. Daniel
367 Lt Col A.W. Hallows OBE
368 Col A.E.W. Stormer
369 Capt R.N. Howard
370 Capt C.C. Lacey
371 Maj W.J. Dixon
372 Maj G.T. Cook

373 Maj C.J. Upchurch
374 PMC, Byams House Officers Mess, Marchwood
375 Maj I.S. Ormerod
376 Capt P.J. Clayton
377 Lt Col G.M.J. Grieve
378 Maj D. Heffron MBE
379 Maj R.A. Pinnock TD
380 Lt Col H.M. Dannatt
381 Lt Col A.F. Baker
382 Capt L.R. Bedborough
383 Capt R.M. Bush
384 Lt Col J.W. Heath
385 Maj K.O. Proudman
386 Lt Col R. Morgan MBE
387 17 Port Regt RCT
388 Brig D.T. Kinnear OBE
389 Lt Col R.E. Ratazzi
390 Col H.V.C. Stephens
391 Lt D.M. Russett
392 Lt Col P.E. Burton
393 Maj N.P. Humpherson
394 Maj V.J. Sanders
395 Maj H. Aldred
396 Maj A.C. MacBeth
397 Maj I. Rouse
398 Lt Col R.E. Thomas OBE
399 Capt R.F. Mutch
400 Capt J.M. Jones
401 Maj P.B.G. Cummings
402 Capt G.F. Charlton
403 Capt A.H. Rashbrook
404 Maj J.M. Bowles MBE
405 Capt M.D. Vickers
406 Maj R.I. Kilpatrick
407 H.Y. Wakeling
408 Capt G.J. Wright
409 Lt Col C.R.C. Green
410 Col I.R. Jones OBE
411 Maj J.B. Shaw
412 Lt Col S.C. Ivatt MBE BEM
413 Lt Col M.B. Hardwick
414 Capt C.R. Prince
415 Capt M.A. Smith
416 Maj S.B. Doudney
417 Lt Col D.S. MacFarlane
418 Lt Col V.H. Vaughan MBE BEM
419 Capt D. Brown
420 156 (Merseyside and Greater Manchester) Tpt Regt RCT(V)
421 Lt Col A.G. Morrison MBE
422 Capt P.W. Bessey
432 Brig J.J. Regan
424 Col H.T. Wheeler ERD
425 Maj P.J.C. Payne
426 Maj G.T.E. Dawe MBE
427 Lt Col K.E.F. Allman TD
428 Lt Col G.M. Sprague
429 Maj P.R. Stabb
430 Maj G. Sharpe
431 Maj S.J. Holman
432 2Lt A.J. Dyer
433 Maj D.W. Dawson
434 Maj A.G. Cross
435 Capt R.E.G. Judge
436 Maj P. Firth
437 Capt R.S. Card
438 Maj P.G. May
439 Lt Col D.L. Davies TD
440 Maj E.G. Waite-Roberts TD
441 Maj A.R. Grimshaw
442 Maj L.R. Eaton MBE
443 Maj D.J.R. Martin
444 Maj L.C.W. Reason
445 Lt Col J.D. Hennell
446 Maj (QM) B.R. Hadlow
447 Brig A.P. Campbell OBE
448 Maj B.V. Wynn-Werninck
449 Lt Col A.R. Price
450 Maj R.H. Woodward TD
451 Maj I.H. Stewart
452 Lt M.R. Israel
453 Brig G.E. Bavin
454 Capt M. Rees
455 Lt Col G.D. Williams
456 Warrant Officers & Sergeants Mess, 3 Armd Div Tpt Regt RCT
457 Officers Mess, 3 Armd Div Tpt Regt RCT
458 Maj D.M. Bond MBE
459 Col B.A. Edridge
460 Col B.J. Harban MBE
461 Maj G.B.L. Fox
462 Maj A.H. Sutton TD
463 Lt P.A. Marshall
464 Capt D.K. Baker
465 Lt W. Robinson
466 Maj S.L. Davies
467 Capt D.I. Hollas
468 Maj J.M. Keir TD
469 Col R.H. Dadswell
470 Maj K.A. Hearson
471 Maj C. Leefe
472 Brig H.R. Orton
473 Capt G.P. Gowlett
474 Capt (QM) T.W.A. McMinn
475 Lt Col D.D. Burns
476 Lt Col M.G.R. Hodson
477 Maj I.W. MacArthur
478 Maj M.G. Roach
479 Officers' Mess, 10 Corps Tpt Regt RCT
480 WO & Sgts Mess, 10 Tpt Regt RCT
481 Maj R.L. Martin
482 Capt G.O.H. Snelgar
483 Capt I.G. McKend
484 Maj G.A. Fitness
485 Capt D. Farquherson
486 Maj J.G. Taylor
487 Mrs S.J. Mead
488 Lt Col J.R. Cawthorne
489 J.E.O. Arnold
490 Maj (QM) E.A. Ford
491 Lt Col A.J. Samson
492 Lt Col D.F.J. Horner
493 Lt Col M. Cowan
494 Lt Col L.C.L. Cook
495 Brig F.A. Shaw OBE
496 497 Maj K.W. Blease
498 500 D. Nicholas
501 26 Sqn RCT
502 Maj A.M. Paterson
503 Maj J.M. Shorer
504 Maj P.B. Wilkie
505 Capt J.W. Capeling
506 Capt L.M. Robinson
507 WO1 K. Dunstan
508 Officers' Mess, 20 Maritime Regt RCT
509 WOs' & Sgts' Mess, 20 Maritime Regt RCT
509 WOs' & Sgts' Mess, 20 Maritime Regt RCT
510 Wardroom Mess, HMAV Ardennes
511 Wardroom Mess, HMAV Arakan
512 Lt Col A.R.N. Jackson
513 Maj R.S. Garrick MBE
514 Maj R.W. Carter
515 Maj G.T. Collinson
516 WO1 R. McPherson
517 518 Capt R.R. McMordie
519 520 Maj M.J. Rust
521 522 WO1 (RSM) B.A. Edwards
523 524 Maj R. Wilkins
525 Lt Col H.C. Attree-Palmer MBE
526 Maj C.R.V. Wallace
527 Maj B.J. Etheridge
528 Capt N.A. Griffin
529 Maj J.H. Stretton
530 Maj P.W. Harper
531 Maj A.W. Floyd-Jewel
532 Lt Col J.M. Wetherley
533 Capt J.D. Murray
534 SSgt R.L. Griffiths
535 WO2 J.R. Low
536 WO1 (RSM) N. Dempsey
537 SSgt A.P. Smith
538 Sgt K. Randles
539 Sgt P. Shoemaker
540 24 Tpt Regt RCT
541 Maj A.J. Hughes
542 Lt Col T.H. Ridgway
543 Maj J.F.E. Burgess
544 Lt Col H.F. Rayner
545 Maj J.R. Wallace
546 Capt J.P. Taylor
547 Brig L.I. Aspland OBE
548 Brig D.W.E. Hancox
549 Lt Col R.F. Grevatte-Ball
550 Col B.E. Colston
551 Lt Col S.B. Ball OBE
552 Maj R.M. Wilkinson TD
553 Maj R.R.C. Cooke
554 WO1 J. Baker BEM
555 WOs' & Sgts' Mess, 151 Tpt Regt RCT(V)
556 WOs' & Sgts' Mess, 210 Sqn RCT(V)
557 WOs' & Sgts' Mess, 215 Sqn RCT(V)
558 WOs' & Sgts' Mess, 240 Sqn RCT(V)
559 WOs' & Sgts' Mess, 562 Sqn RCT(V)
560 WO2 K. O'Leary
561 WO2 D.D. Davis
562 SSgt W.A. Proctor
563 Maj A.D.F. MacLean TD
564 W.H. Earl
565 Lt Col R.E. Wills
566 Lt Col B.H.R. Coles
567 Lt Col R.E. Collins
568 Lt Col W.B. Gray
569 2Lt A.J. Gale
570 Lt Col M.A. Ponikowski
571 572 A.A. Orgill
573 Thomas R. Whitley
574 Thomas E. Bell

575 Stanley Johnstone Little
576 A.C. Crowson
577 Charles C. Stadden
578 Brian Frederick Little
579 Lt (QM) M. Williamson
580 S. Binney BEM
581
582 J.C. Holliman
583 P.G. Craven
584
585 Maj W.R.A. Selbie
586
590 Maj (QM) A.D. Blyth
591 P.G. Cole
592 Alec Brown
593 James A. Murdoch
594 Ronald Hogg
595 F.J. Heathcote
596 Lt Col Wm C. Munro OBE TD
597 Capt W.J. Whitty TD
598 Maj D.J. McCarthy
599 Maj L.R. Tucker
600 Maj R.L.A. Bancroft
601 Maj P.C. Marsden
602 Lt Col G.W. Somerville
603 Maj I.W. Abbott
604 Lt Col J.A.I. Baldwin
605 Capt C.R.G. Davies
606 Rev Graham Thorne
607 Lt Col W.A. Lyons BEM
608 Capt K.G.H. Brook
609 Maj A.J. Barnard
610 Lt Col D.A.G. Welsford
611 Capt K.B. Carey
612 B.W.P. Egerton
613 WO2 D.V. Hill
614 John W. Reid
615 Eric A. Anderson
616 Col E.A. Lewis
617 Sidney W.R. Kent
618 F.C. Butcher
619 Capt (QM) I. Streeter
620 Lt Col H.H. Cail
621 D.J. Wride
622 A.R Cantwell MA
623 Lt Col H.M.J. Jensen OBE
624 Andrew B. Chisholm
625 Lt Col W.R. Bloxwich
626 Capt M.J. Underwood
627 Lt Col A.J.C. Fisher MBE
628 SSgt M.R. Blackman
629 Lincolnshire Library Service
630 Imperial War Museum
631 Lt Col M.J. Bennetts
632 15 Sqn RCT
633 Maj M.J. Paine
634 Maj. M.R.P. James
635 Maj A.L. Jones
636 F.K.B. Benson
637 P.N. Murphy
638 F.F.A. Nunez
639 John Abbott
640 Kenneth Albert Holland
641 WO2 (RQMS) D.J. Griffin
642 Lt Col A.G. Bisset
643 T/112482 Nicholson T. Ex 503 Coy RASC
644 Kenneth W.S. Goodson
645 Sydney J. Caddell
646 Christopher Thomas
647 East Sussex County Library
648 Maj P.H. Wood
649 WO1 K.J. Beckley
650 Col H. Muddiman OBE

Remaining names unlisted.

ENDPAPERS: Places where the RASC served, 1919-1939.

Places where the R